SpringerBriefs in Earth Sciences

SpringerBriefs in Earth Sciences present concise summaries of cutting-edge research and practical applications in all research areas across earth sciences. It publishes peer-reviewed monographs under the editorial supervision of an international advisory board with the aim to publish 8 to 12 weeks after acceptance. Featuring compact volumes of 50 to 125 pages (approx. 20,000–70,000 words), the series covers a range of content from professional to academic such as:

- timely reports of state-of-the art analytical techniques
- bridges between new research results
- snapshots of hot and/or emerging topics
- literature reviews
- in-depth case studies

Briefs will be published as part of Springer's eBook collection, with millions of users worldwide. In addition, Briefs will be available for individual print and electronic purchase. Briefs are characterized by fast, global electronic dissemination, standard publishing contracts, easy-to-use manuscript preparation and formatting guidelines, and expedited production schedules.

Both solicited and unsolicited manuscripts are considered for publication in this series.

More information about this series at http://www.springer.com/series/8897

Juan Carlos Mora Chaparro

Mapping the Risk of Flood, Mass Movement and Local Subsidence

A New Proposal for Major Cities

Springer

Juan Carlos Mora Chaparro
Universidad Nacional Autónoma de México,
Instituto de Geofísica, Campus Michoacán
Mexico City, Mexico

ISSN 2191-5369 ISSN 2191-5377 (electronic)
SpringerBriefs in Earth Sciences
ISBN 978-3-030-22471-4 ISBN 978-3-030-22472-1 (eBook)
https://doi.org/10.1007/978-3-030-22472-1

This Springer imprint is published by the registered company Springer Nature Switzerland AG
The registered company address is: Gewerbestrasse 11, 6330 Cham, Switzerland

MAPPING THE RISK OF FLOOD, MASS MOVEMENT AND LOCAL SUBSIDENCE

A NEW PROPOSAL for major cities.

Preface

This book describes three techniques to identify and map the risk associated with flood, mass movement and local subsidence phenomena in urban areas and along communication routes. Diagrams and photographs enable users to relate and apply the findings to their own locality to develop preventive programmes and minimise social vulnerability, and pinpoint sites for works to reduce or eliminate exposure to danger. The techniques are intended mainly for urban areas where there have already been problems and where it is likely is that these will recur to cause the same or greater damage.

The book is divided into three parts:

Chapter 1. Identifying and Mapping the Risk of Flood in Urban Areas.
Chapter 2. Identifying and Mapping the Risk of Rockfall and Landslide on Roads and Urban Areas.
Chapter 3. Identifying and Mapping the Risk of Local Subsidence.

Each method is intended to be highly practical, simple and easy to use. In fact, many of the activities involved are undertaken already by civil protection personnel during an emergency, for example cordoning off the zone in question. What is proposed is that these actions are carried out methodically and with up-to-the-minute data so that they complement each other to update information reliably.

This book is aimed at civil protection personnel, civil defence professionals and decision-makers concerned with risk management, growth and urban planning. Its main objective is to be a useful tool for those staff with basic knowledge who are daily exposed to various types of risk in urban areas and along roads, and to facilitate schemes for preventive work. It is the first book to attempt, to identify and map the social construction of risk alongside their vulnerability, both physical and social, to flood, mass movement and local subsidence in these areas.

Mexico City, Mexico Juan Carlos Mora Chaparro

Acknowledgements

Design—D. C. G. Paola Lisandra Ramírez Pérez
Style correction—L. C. P. and A. P. Luz María Ruiz Pelcastre

Introduction

Research into the risks associated with natural phenomena has gained an importance due to the drastic increase in these events and their repercussions on the population and economy. For this reason, various organizations carry out investigations on the theoretical-methodological elements to build a consensus on definitions and methods for the analysis and quantification of danger, vulnerability and risk. These organizations include the National Center for Disaster Prevention (CENAPRED, Mexico), the Economic Commission for Latin America and the Caribbean (ECLAC), the Network of Social Studies in Disaster Prevention of Latin America (LA RED). They aim to reduce the exposure to risk for the human population.

Despite some progress, due to their complexity, not all phenomena of either natural or anthropogenic origin have a specific means of analysis. These include flood, landslides and local subsidence in urban areas. For their analysis, independent research is needed in order to offer new ideas and reasoning, using novel theoretical and methodological elements, to widen the investigations into hazard and vulnerability.

Warning: It is the responsibility of the user, when using the techniques presented here, to take all necessary preventive measures when traversing active areas of subsidence, landslide, rockfall and flood, since failure to do so may expose them to danger.

Contents

List of Figures

List of Tables

Chapter 1
Identifying and Mapping the Risk of Flood in Urban Areas

J. C. Mora Chaparro, *Mapping the Risk of Flood, Mass Movement and Local Subsidence*, SpringerBriefs in Earth Sciences, https://doi.org/10.1007/978-3-030-22472-1_1

1.1 Introduction

Year after year, many urban areas experience flooding and parts that did not have the problem before are newly affected. The main reason for new floods is changes in land use associated with construction due to accelerated urban growth affecting existing localities. Population growth in cities requires new housing, and this development demands increased drainage networks, drinking water and roads. As the urban growth is not planned, connecting new drains to those built initially swells the flow to the extent that it often exceeds the intended capacity, so that in the rainy season water may be seen emerging from sewers and drains, not being absorbed and taken away. In the same way, when new roads are connected to old ones, the extent of the hard, impermeable run-off area is increased by the asphalt, the length of the roads and, therefore, the amount of surface water flow. Rainfall becomes run-off that flows down to lower areas, causing ponding and consequently flood. This is because there is no urban planning to consider growth and prepare the drains and roads built in earlier times to accommodate the new development.

Constructing and connecting hard-surfaced roads in an urban area creates an urban watershed in which the roads turn into urban tributaries that carry water to a main urban watercourse. A natural hydrological basin is bounded by its watershed, wherein various micro-watersheds with tributaries discharge into a main river that takes it out of the basin, or into a lake or lagoon. When an area becomes an urban watershed due to development, it is useful to delimit the various micro-watersheds in the same way, with their urban tributaries, discharging into a main watercourse or a flood zone. Often, the water from natural tributaries passes beneath this urban system in culverts, or joins the flow along the road system to become an urban tributary.

The proposed technique makes it clear that an area that has already flooded in the rainy season is vulnerable to the same or greater damage after a similar intensity of rain. Within the flood hazard zone, there are three different levels of danger:

(a) High. The height of floodwater is as much as or greater than 10 cm, and inhabitants' furniture, electronic devices and daily activities are affected.
(b) Medium. There is a depth of between 5 and 10 cm of floodwater.
(c) Low. The floodwater is between 1 and 5 cm.

Certainly, many researchers engaged in the analysis of flood risk recognise the different severity levels. However, as mentioned above, the classification offered here takes into consideration the views of those affected by flood, established by means of 1230 surveys, from SEDATU 2012 and 2015, during the following floods: Tuxtla Gutiérrez (2015), Tapachula and Suchiate (2012) Risk Atlas, Chiapas; Xochimilco (2012), Milpa Alta (2012) and Tlahuác (2012), Mexico City; Nezahualcoyotl (2015), State of Mexico Atlas.

The vulnerability within those areas known to be floodplains was determined both by the extent of the damage to housing and infrastructure (physical vulnerability) and by the accounts of the inhabitants or users (social vulnerability).

The outcomes from the application of this technique are:

- Maps of high-, medium- and low-hazard areas.
- Maps of physical vulnerability; these assess the extent of the housing and infrastructure in each area at each of the three levels of flood risk.
- Social vulnerability maps, in which the numbers of inhabitants and users of infrastructure in the areas at the three levels of flood risk are estimated.
- Risk maps, defined by overlaying hazard maps and vulnerability maps at each of the three levels of risk.
- Map of the urban watershed in question and its sub-basins, identifying the main urban watercourse.

This technique can act as a tool for civil protection personnel, public works and authorities across the country's states, communities, municipalities and districts that experience flooding.

At present, during the response to a flood, 60% of the procedures involved in the proposed technique are already in use. The flood area is cordoned off (hazard zone), housing and people are counted (assessment of physical and social vulnerability) and, finally, there may be an evacuation. By collecting data during the response to an emergency and using the proposed procedures, the areas that bring the water to the affected area (urban watersheds, hydrographic sub-basins and urban tributaries) can be identified, and consequently their origin. By relating each sub-basin to precipitation records, the volume of the water brought down can be calculated. In this way, the areas that contribute the greatest amounts of floodwater can be identified and thus preventive programmes or works proposed to eliminate or diminish the danger.

With this technique, it is possible to establish what reduces the risk of flood, since identifying those areas sending the greatest volume of water through their tributaries allows the necessary projections to be modelled. In addition, accurate data will be available when developing urban growth so that the new drainage connections do not exceed the volume that can be drained or is able to reach the sewage system and thus generate new flood-prone areas.

Without any doubt, the areas that do not suffer from flood are those that are well planned and whose hydraulic networks cope with the flow, while in the areas in danger of flood the capacity of the hydraulic network has been exceeded, causing water build up.

1.2 Background

Numerous studies have been carried out to understand the problem of flood, and various methods have been proposed to map the danger, vulnerability and risk in urban areas.

Bertoni (2004), cites the two main types of factors of flooding in urban areas:

(a) riparian flooding, generally associated with undue urbanization of the floodable areas adjacent to watercourses
(b) floods caused by traditional urban growth.

CENAPRED (2006) notes that the danger of flooding is based on certain criteria, such as the height of ponded water and the speed of the flow in roads, rivers and streams. When a river's natural course overflows, it is known as a fluvial flood; when rainfall in an urban area continues after total saturation of the soil, excess water will stay on the surface for hours or days, known as rainfall flooding. Other phenomena can cause a flood, such as high tides, where the normal sea levels are exceeded, producing coastal flooding, tsunamis or the failure of hydraulic works.

Mobayed (2009) presents a flood simulator for urban areas for forecasting purposes and notes that its model of risk exposure is based on the hydrological principle of transit (model 1D) through a simplified form of the Saint-Venant equations (convection-diffusion model). The results are just an approximation of the phenomenon, not only because of the hypotheses used but because of the approximate physiographic representation of the city and the distribution of precipitation data.

Alcocer-Yamanaka et al. (2012) mention that implementation of a rainwater management programme has four phases: identification of the problems caused by flood; hydraulic and hydrological analysis of the area; selection of appropriate tools to build a simulation; and analysis of flood scenarios for various recurrence periods, thus generating possible solutions to mitigate the problem.

Rodríguez et al. (2014) developed mathematical models to determine flood hazard, vulnerability and risk maps, and applied them to the city of Tuxtla Gutiérrez, Chiapas. Likewise, Mora et al. (2015a, b) analysed flood polygons and urban hydrographic basins in preparing the *Atlas of Risks* for Xochimilco and Tlalpan, Mexico City. In contrast, Yamanaka et al. (2016) compiled and analysed various criteria at international level, considering water tension (level), speed, roll-over effects and landslides generated by run-off, achieving hazard maps closer to reality. The same authors mention that both hazard and vulnerability maps should be used to construct risk maps to enable quantification of probable damage in various recurrence periods.

Some of the main reasons behind newly flood-prone urban areas are exponential population growth and poor planning for urban development, because the drainage services are designed to accommodate a specific level of urbanization and inhabitation, and this has been exceeded. Although many methods have been suggested, the problem remains and is becoming acute. For this reason, the proposed technique is for areas that already have a flooding problem, in order to show the steps to take to find its cause. In this way, the authorities can devise preventive programmes or reduce the damage, and can develop a growth plan that does not contribute to new and greater flood.

1.3 Transformation of a Natural Watershed to an Urban Watershed

A natural watershed is defined as an area in which water flows from higher to lower parts into a main watercourse (Fig. 1.1).

In a natural watershed, water flows from the upper parts to a main watercourse along tributaries. Over time, soil, forest, crops and biodiversity develop and communities grow up (Fig. 1.2a). The presence of water is a characteristic of urbanisation, and in the lower parts of a natural watershed there is greater development due to the flatter land and fertile soils there (Fig. 1.2b).

The valleys feeding into the main river undergo change when passing through urban areas because of a reduction in their hydraulic area due to both impermeable surfaces and culverts that pass under houses and streets, whether in pipes and vaults or merely underground (Fig. 1.3).

Urban development in the lower part of a natural watershed follows the natural topography; that is, the gradient of the channels of tributaries, roads and drains that discharge into the main watercourse, which in most cities passes through the centre (Fig. 1.4).

Under urbanization, a natural watershed becomes an urban watershed, and the roads turn into urban tributaries discharging into a flood zone or main watercourse (Fig. 1.5). It is now a delimited urban watershed with urban tributaries. Rainwater starts to run off and will follow the steepest gradient down roads. It is in equilibrium

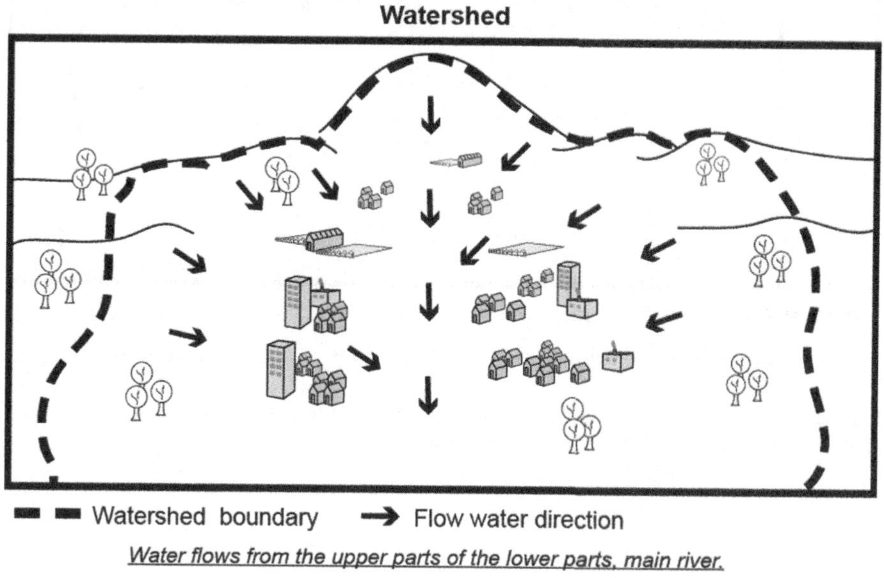

Fig. 1.1 Watershed and direction of currents. The watershed delimits the basins through which water drains to a main watercourse

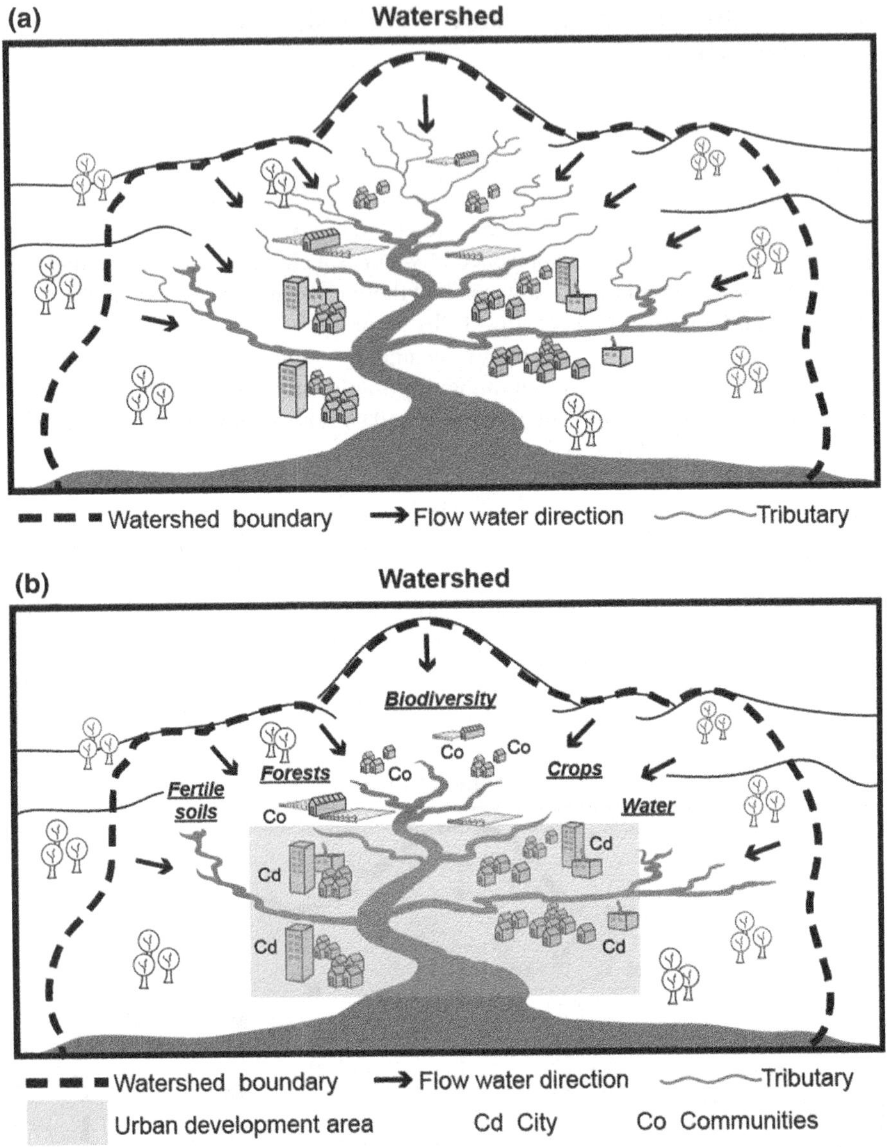

Fig. 1.2 **a** Components of a natural watershed tributary, main river and water. **b** Natural watershed, with development of soil, forest, communities and cities

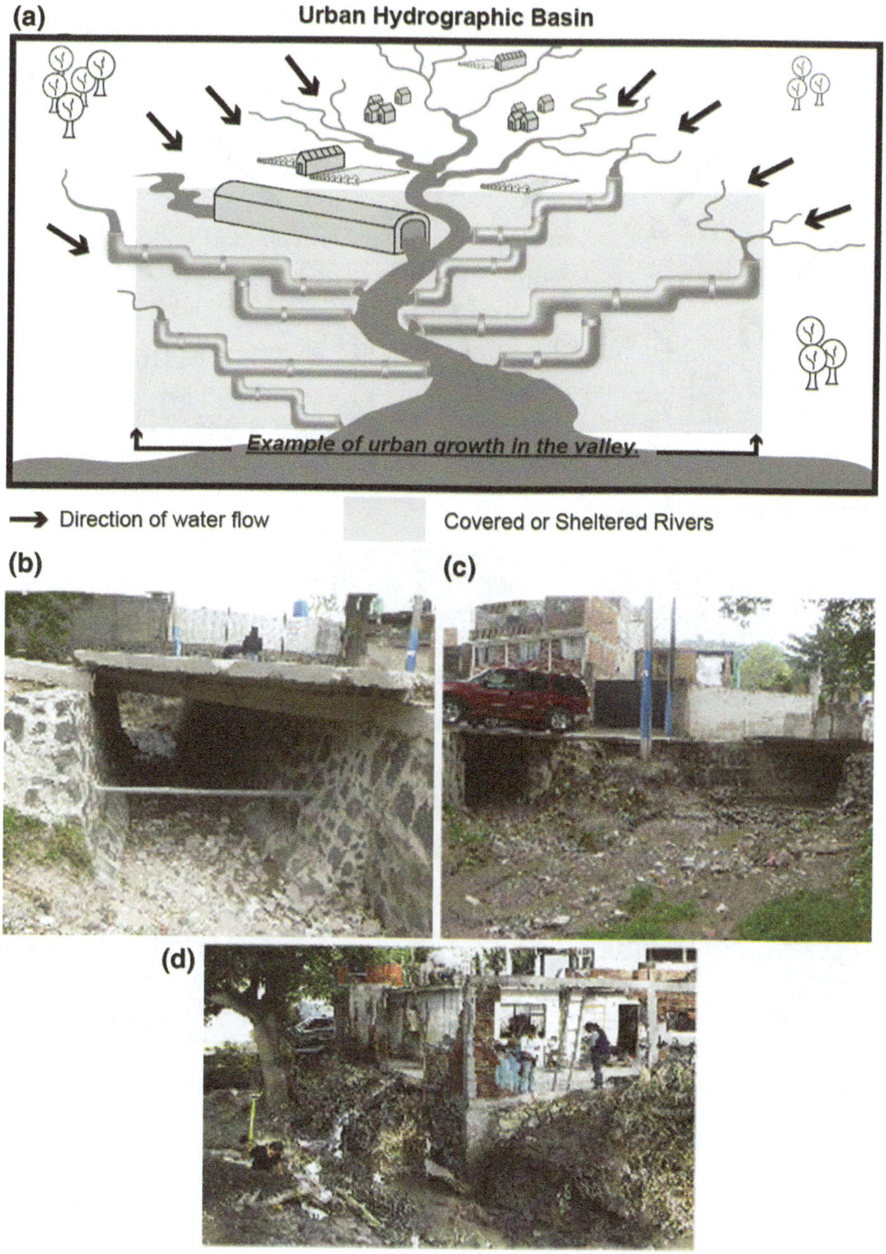

Fig. 1.3 **a** Development of city over tributaries of main river; **b** and **c** The beds are cased or vaulted below road level; **d** A tributary swept away a building during a flood and deposited transported material, since it became dammed and had greater force owing to the reduction in its hydraulic area

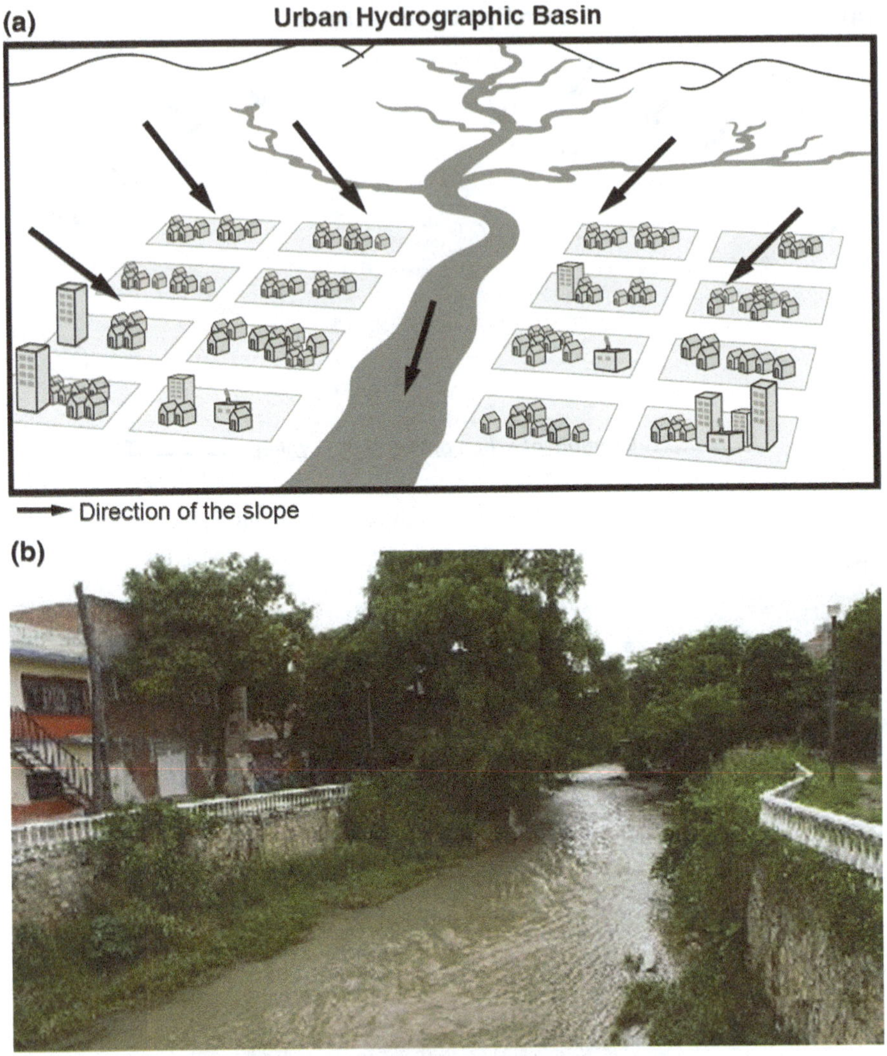

Fig. 1.4 **a** Urban development, with the direction of the original gradient (arrowed) towards the main channel through the city. **b** Sabinal River in Tuxtla Gutiérrez: the riverbanks are now covered by masonry and form the basis of roads and housing

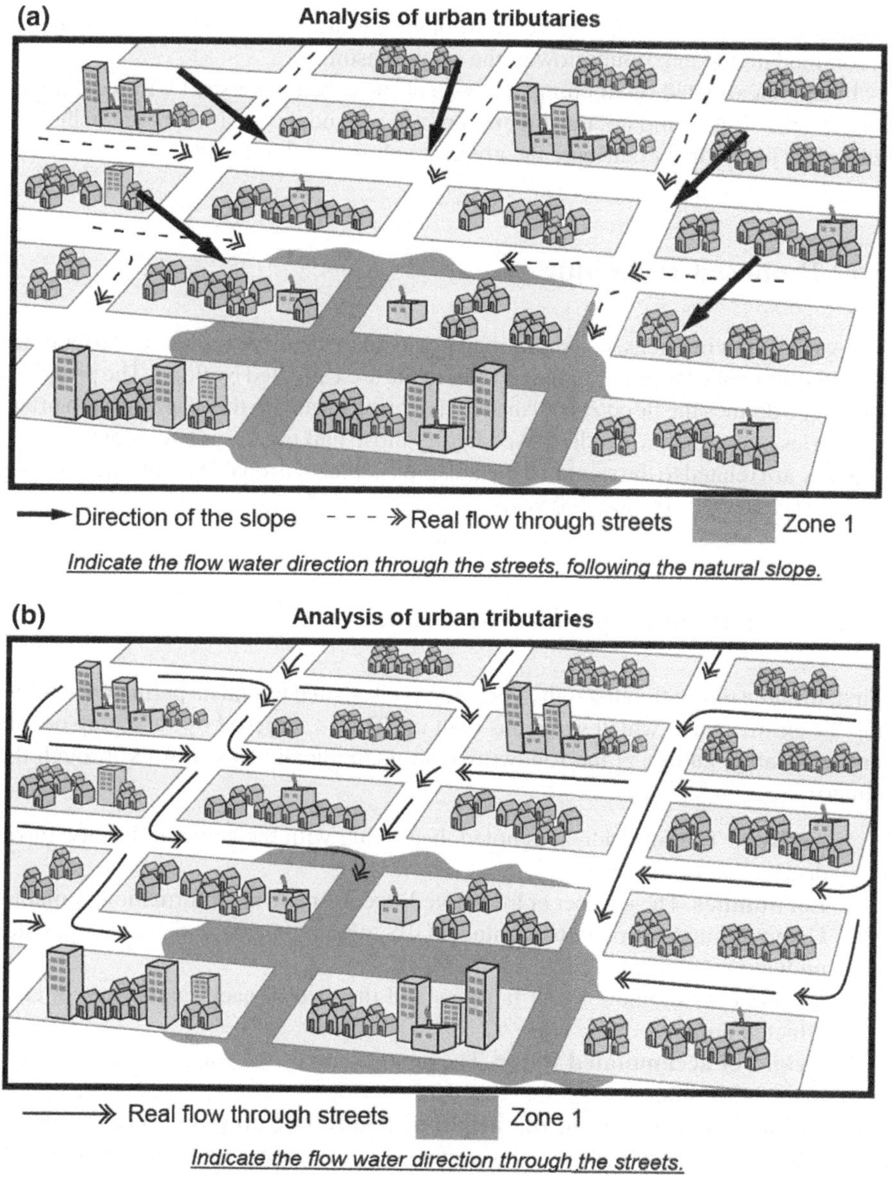

Fig. 1.5 **a** and **b** Direction of the slope and the actual direction that the water flows in each of the roads into a flood zone (Zone 1)

when most of its rainwater is filtered through the drains and sewers, calculated to accommodate the maximum flow in the rainy season.

In an area affected by flooding, analysis of the urban watershed will indicate the cause of the water build up, the direction of run off and the areas (sub-watersheds or sub-urban basin) that contribute the greatest volume of floodwater.

1.4 Proposed Technique

Using the following steps, it is simple and practical to identify the areas and tributaries that contribute the largest volume of water to the area affected by flood. The proposed technique defines the flood zone. An area that brings water to this zone through urban tributaries is an urban watershed. We identify those making the greatest contribution, and they are related to the rainfall data and, if possible, to probable recurrence periods of 2, 5, 10, 20, 50, 100 and 200 years.

1.4.1 Delimitation of Flood Risk Zone

First, the area that is flooded—the flood zone—is visited to find its perimeter, enclosing the entire area in which water has built up. This is achieved by surveying people in this locality, gathering the following data methodically as you walk through the streets:

1. **Address**. The full address is noted, beginning with the name of the city, town, district and street details.
2. **Lot number**. The number or lot is noted, according to the information available.
3. **Geographical coordinates**. Using a GPS, obtain the geographical coordinates midway through each batch.
4. **Members**. Note how many families and the size of each family live in each affected home.
5. **Height of accumulated water**. The height (cm) of the water build-up in front of each lot is noted.
6. **Record any evidence of the height of floodwater on poles, posts, walls or other**. When walking through the streets, photographic and numerical records are made of the height of water on poles, walls, trees, etc., if any.
7. **Record data for all lots**. Data should be recorded for each lot, even if the residents/users do not answer or do not undertake the survey, since the average number of people per lot is calculated later.
8. **Walk through all the streets in the flood zone**. Every road in the flood-affected area where water has ponded up should be visited.
9. **Gather reliable data**. If is not possible to record all the lots within the flood zone, 80% is acceptable.

10. **Calculate the average number of inhabitants per lot.** Using the data obtained from surveys, extrapolate to find the average value for those lots whose residents did not undertake the survey (Fig. 1.6).

1.4.1.1 Information Processing

Identify the level of danger.

1. The height of the water build up in front of each lot on the street plan should be recorded (Fig. 1.7).
2. Points with of the same level of water build up (isolines) across the whole zone (Fig. 1.8).
3. Highlight or embolden the major isolines, i.e. 0 cm, 5 cm and 10 cm.
4. Colour the area inside the 0–5 cm lines yellow, that inside the 5–0 cm line orange; and that inside the 10 cm line red. In this way, the areas most affected by water build up are most visible.
5. Label the yellow area 'Low hazard <5 cm'; the orange area 'Medium hazard 5–10 cm'; and the red area 'High hazard >10 cm' (Fig. 1.9).

Record the exposure to physical and social risk at each level of hazard.

Each hazard zone is unique, and all lots in the flooded area are counted, as well as the number of lots in the low, medium and high hazard zones. Subsequently, the number of inhabitants in each lot, and in all lots designated as either high, medium or low hazard, is calculated (Fig. 1.10).

By marking out the danger areas seen from the isohyet maps, the number of structures in each and the people inhabiting or using them, according to land use, are ascertained.

During fieldwork, the type of construction or lot can easily be verified from the facade; however, counting the population can only be done by surveying each lot and, on occasion, this will not be possible. When field data are unavailable, calculate the average number of users and inhabitants per lot from the average data obtained from surveyed lots. The result is applied only to lots where no survey was carried out, to achieve realistic, calculated data on the inhabitants of lots within each hazard area. This will give us realistic figures (Table 1.1).

Record of precipitation

The water that builds up in a hazard zone is the result of precipitation running down roads until it accumulates, so it is essential that rainfall records are obtained for the days on and before the flood. A correlation is made between the millimeters of precipitation recorded, the flooded area and the maximum height of water build up. For example, if 170 mm was recorded on the day of a flood, if it rains the same or more in future it may cause similar or greater damage.

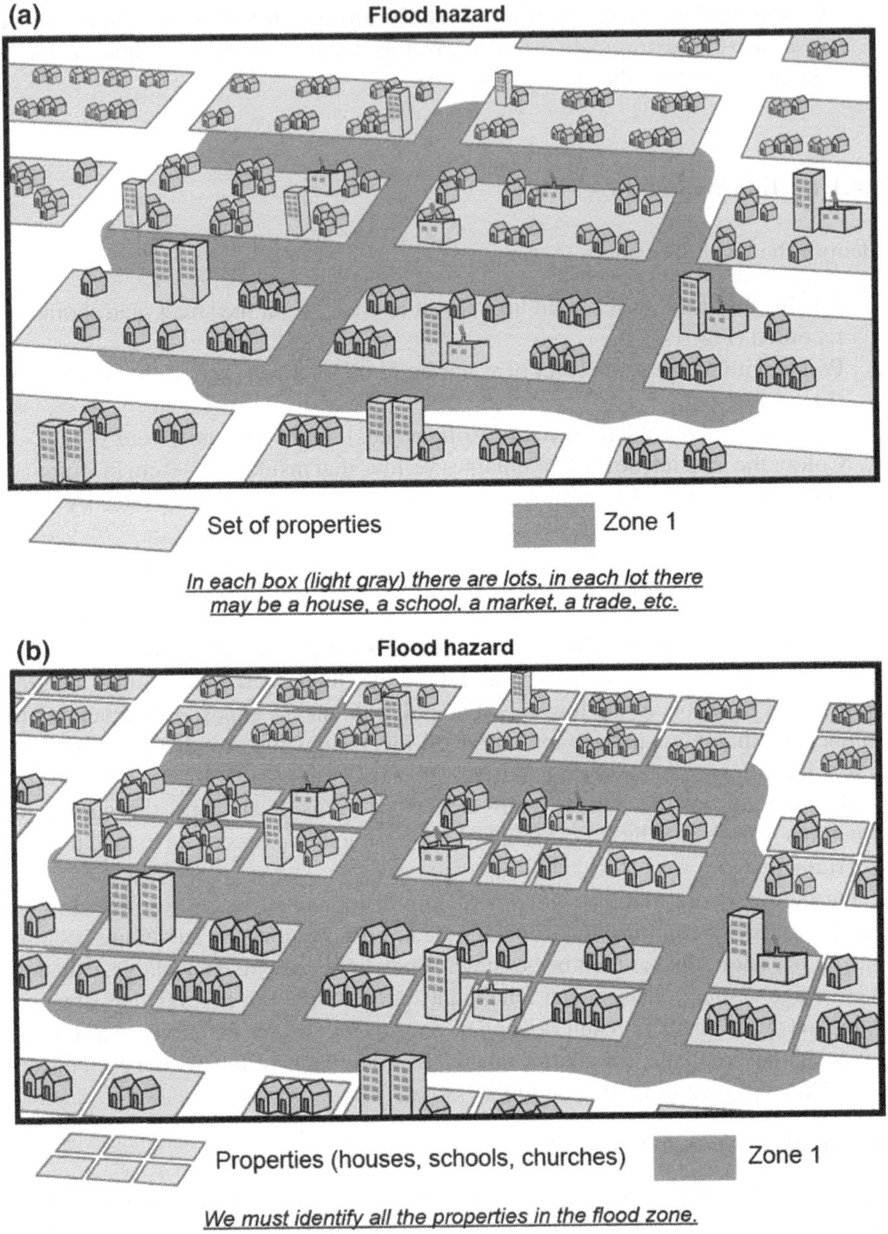

Fig. 1.6 Street plan of division into lots and housings. The grey shading indicates the area affected by flood

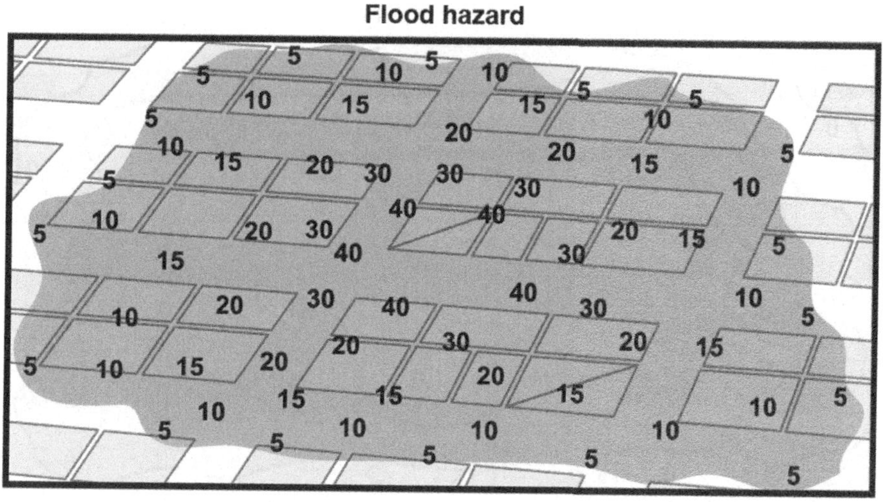

Fig. 1.7 Extent of the flood zone showing data on height of water build up in front of each lot, as obtained from surveying the families living in the lots and measuring of the height of the water stain on walls, posts, trees, etc.

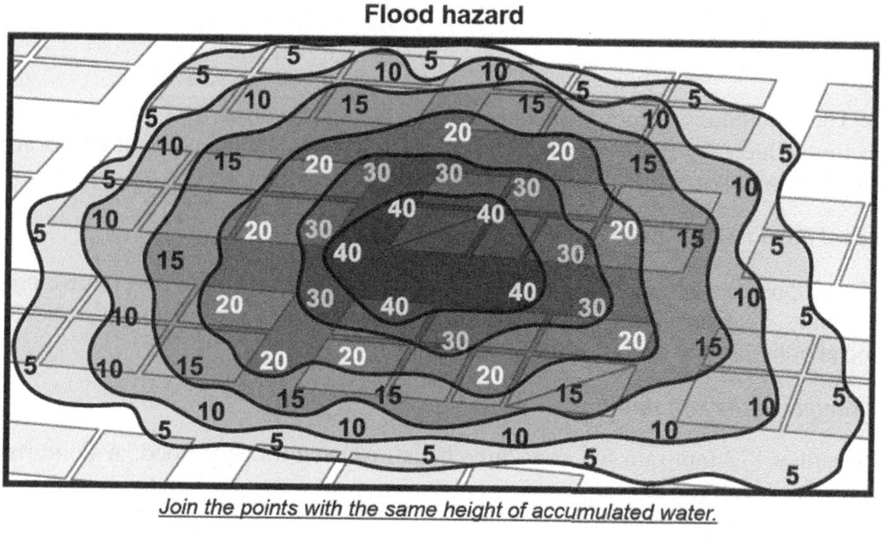

Fig. 1.8 Join up points with the same build up of water in front of each lot. The lines separate intermediate values: for example, between two points A = 5 and B = 20 are areas with 10–15 cm

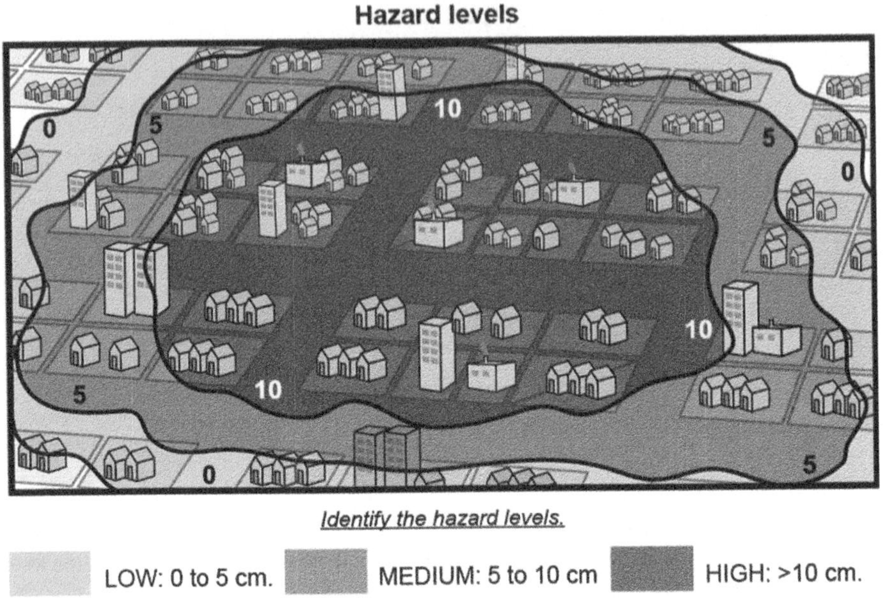

Fig. 1.9 Areas with a water buildup of 0 cm–5 cm (Low Hazard); 5 cm–10 cm (Medium Hazard) and >10 cm (High Hazard)

The precipitation records should be local, as at each urban site there are many factors influencing drainage and build up, such as poor maintenance of drains and construction of new infrastructure and new urban areas.

Local flood risk assessment

The proposed technique represents an effective tool for developing preventive programmes, as it accurately defines the various levels of danger and the physical and social vulnerability at each level.

Example of the risk opinion

Yesterday 172 mm rain fell over three hours, causing a major flood. A flood-risk zone was identified at:

- **High risk** (>10 cm water accumulation) 56 buildings and 448 people.
- **Average risk** (5–10 cm of water accumulation) 62 buildings and 496 people.
- **Low risk** (<5 cm of water accumulation) 66 buildings and 528 people.

In total, there are 184 buildings and 1473 people at risk of total.

The risks are now identified and, with it, the chance to generate preventive programmes, as well as information on where to locate flood shelters, knowing the exact number of people to be received and the materials required to meet their basic needs.

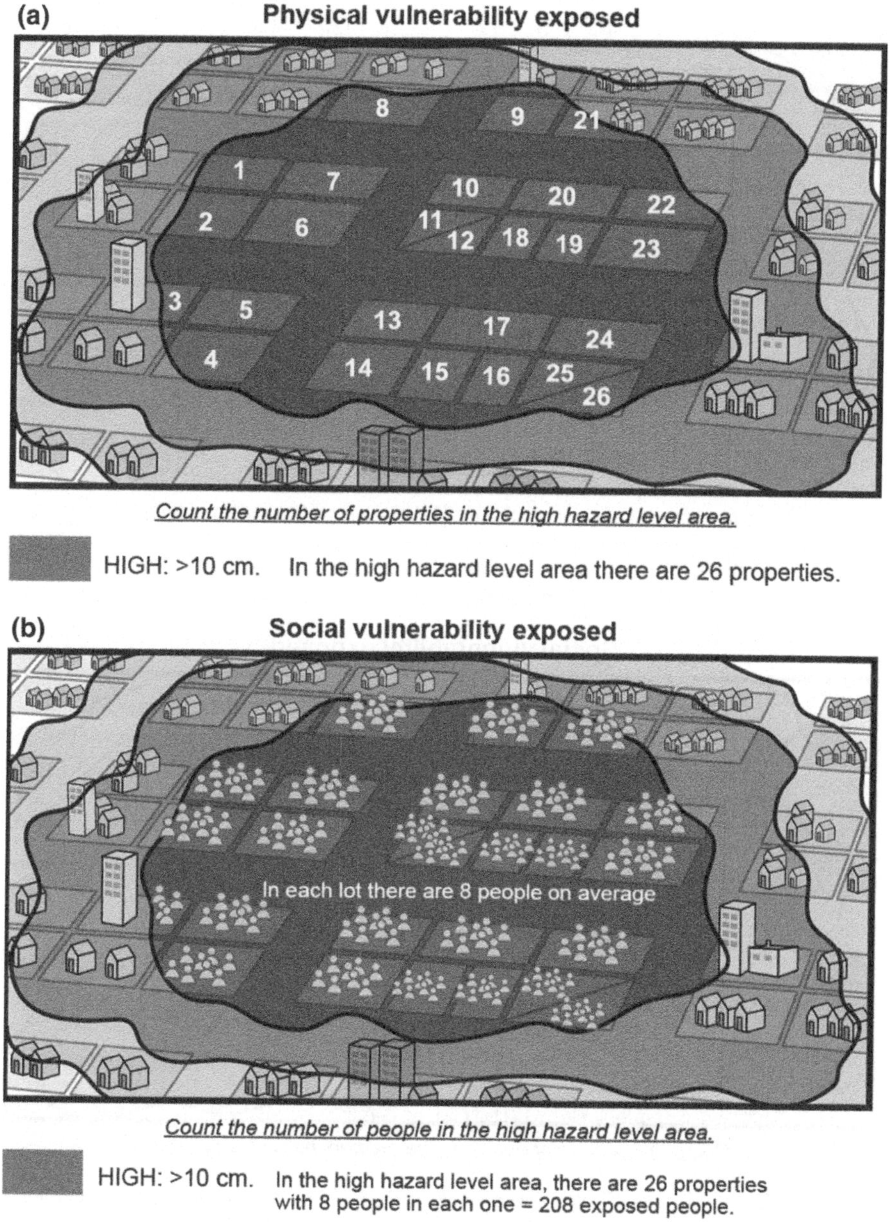

Fig. 1.10 Physical and social vulnerability in the high (**a** and **b**), medium (**c** and **d**) and low (**e** and **f**) hazard zones. Household data are obtained by counting the lots in each hazard area, while social vulnerability is obtained from the calculated average of 8 inhabitants per lot multiplied by the number of lots

(c) **Physical vulnerability exposed**

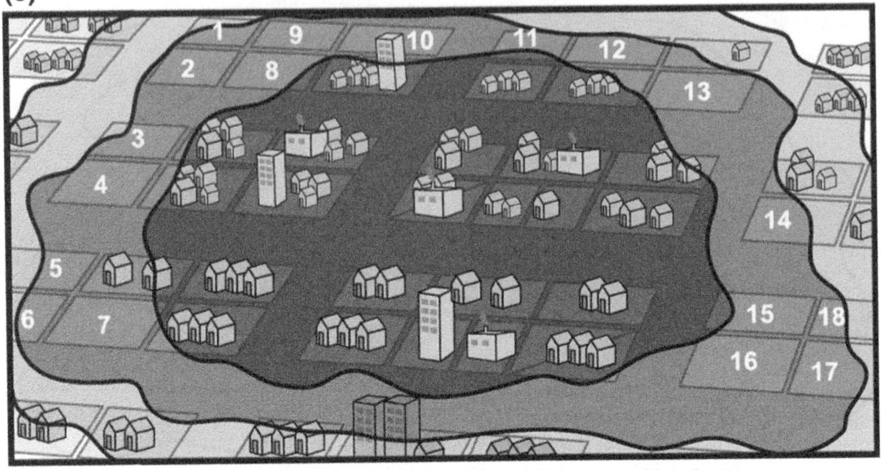

Count the number of properties in the medium hazard level area.

MEDIUM: 5 to 10 cm. In the medium hazard level area there are 18 properties.

(d) **Social vulnerability exposed**

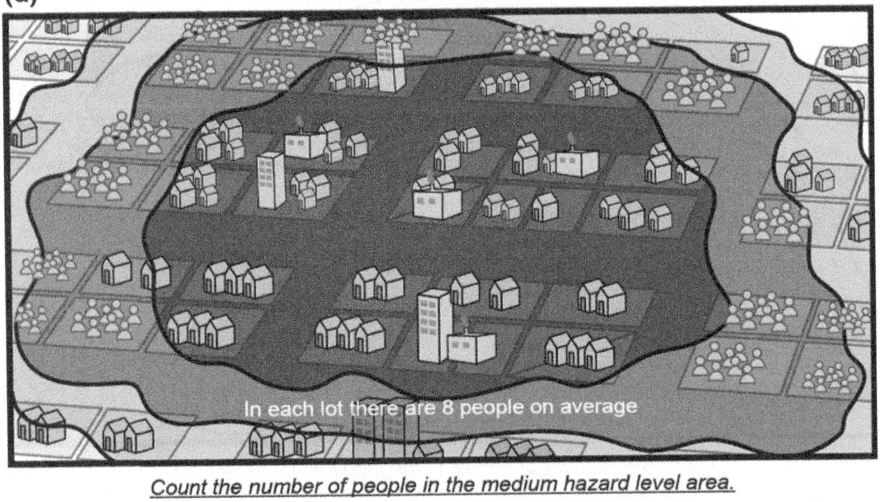

Count the number of people in the medium hazard level area.

MEDIUM: 5 to 10 cm. In the medium hazard level area, there are 18 properties with 8 people in each one = 144 exposed people.

Fig. 1.10 (continued)

(e) **Physical vulnerability exposed**

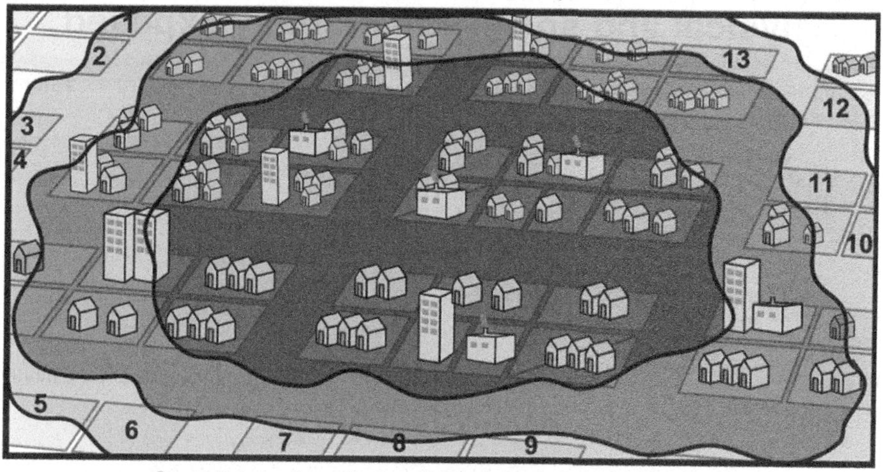

Count the number of properties in the low hazard level area.

LOW: 0 to 5 cm. In the low hazard level area there are 13 properties.

(f) **Social vulnerability exposed**

Count the number of people in the low hazard level area.

LOW: 0 to 5 cm. In the low hazard level area, there are 13 properties with 8 people in each one = 104 exposed people.

Fig. 1.10 (continued)

Table 1.1 Physical and social vulnerability in areas of high, medium and low danger

Physical and social vulnerability exposed

Vulnerability / Hazard	High Hazard	Medium Hazard	Low Hazard	Total
Physical	26	18	13	57
Social	208	144	104	456

Physical vulnerability is assessed on the basis of fieldwork and the street plan. Social vulnerability is assessed on the basis of an average of eight people per lot

1.4.2 2nd Stage. Mapping the Urban Hydrographic Basin, Urban Hydrographic Sub-Basin and Urban Tributaries

Once the flood-risk zones have been mapped, it is essential to identify where the floodwater is coming from. To do this, you need a local street plan that covers the hazard zone, showing all the surrounding roads. Using this plan, walk along down every street in the entire hazard zone.

Fieldwork

1. Walk through the streets and identify where water begins to flow into the flood zone. As you walk, you will be able to identify exactly where water begins to run towards the flood zone. Once you have marked on the map all the points where surface water begins to flow, joining them up shows the area that sends water to the flood zone, delimiting the urban hydrographic basin.
2. All streets within the urban hydrographic basin must be walked through in the rainy season to note the direction in which the water flows to the flood zone. When doing so, note on the sketch or street plan the direction and junctions by arrows to identify the urban tributaries that send water to the flood zone.

Steps to follow

1. Identify the urban hydrographic basin: Join all points from which water flows to map the urban hydrographic basin that sends water to the flood zone (Fig. 1.11).
2. Direction of flow: Mark on a street plan the direction of flow of the urban tributaries towards the flood zone (Fig. 1.11).
3. Classify the level of urban tributary: At street level, classification is carried out as follows. When water starts to run down a road, it is identified as Level 1. When two tributaries join that are classified with the same number, the classification of the road increases in value by 1; for instance, if two tributaries of Level 1 join,

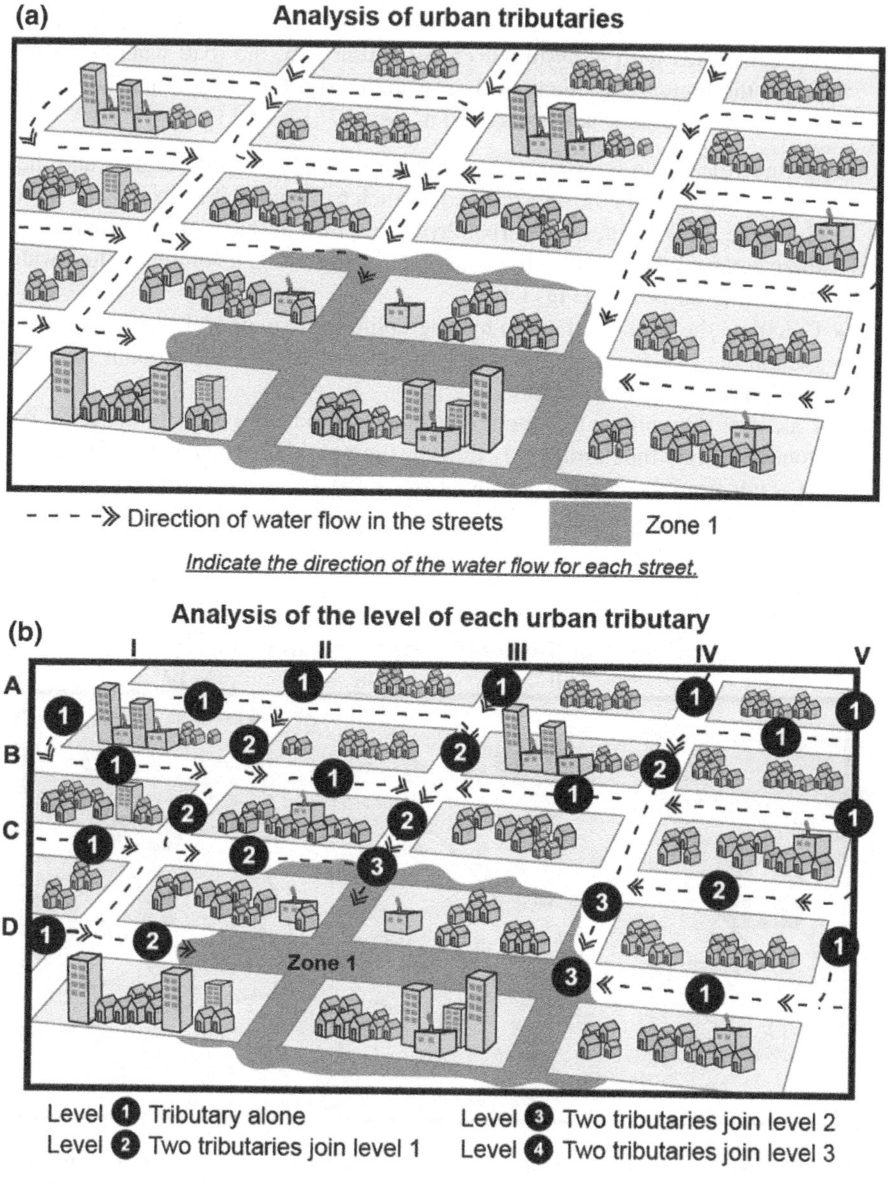

Fig. 1.11 **a** Urban hydrographic basin showing urban watershed (dashed line), urban tributary and the direction of flow (dashed arrows). **b** Classification of each of tributary; Level 1: run-off begins; Level 2: water is carried from two Level 1 tributaries or from one Level 1 and one Level 2; Level 3: water is carried from two Level 2 tributaries or from one Level 2 and one Level 3

its value becomes 2; if two of Level 2 join, it becomes 3, and so on. When two urban tributaries with a dissimilar classification join, the road where they join takes on the higher classification. For example, if a Level 2 tributary of joins a Level 3 tributary, the road is classified as Level 3 (Fig. 1.11).

- Delimit the urban hydrographic sub-basins: Analyze all tributaries and their unions, and identify the area drained by each tributary that leads into the flood-risk zone. This defines each urban hydrographic sub-basin. In an urban basin, there can be two or more sub-basins, each contributing a volume of floodwater depending on its size (Fig. 1.12).
- Calculate the area of all urban hydrographic sub-basins (A1, A2 and A3).
- Obtain data on precipitation: Look up the records for the amount of recent rainfall, for example 172 mm.
- Calculate the volume of water contributed by the sub-basins to the flood risk zone: Multiply the rainfall (172 mm) by their number. The result is the volume of water from the entire watershed (Fig. 1.13). Once the volume of water that is contributed by the sub-basins is calculated, they may be listed in decreasing order of volume, and thus importance.

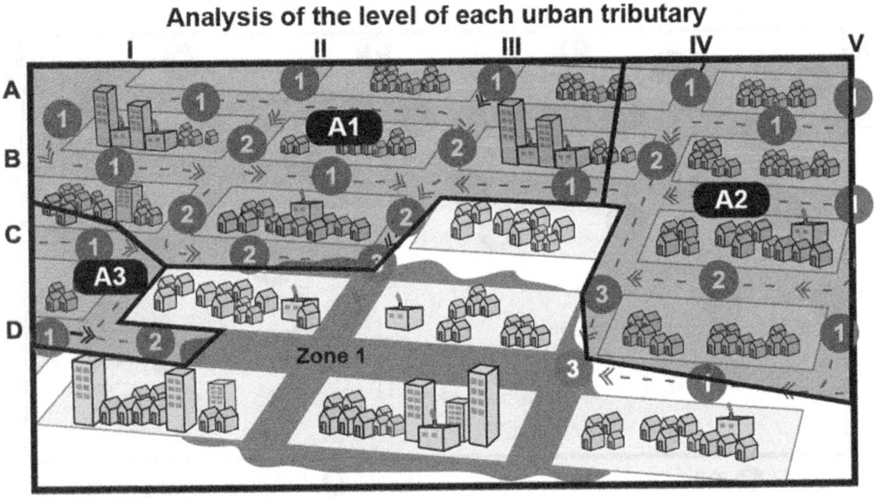

Fig. 1.12 Urban hydrographic sub-basins (A1, A2 and A3) within the main urban hydrographic basin

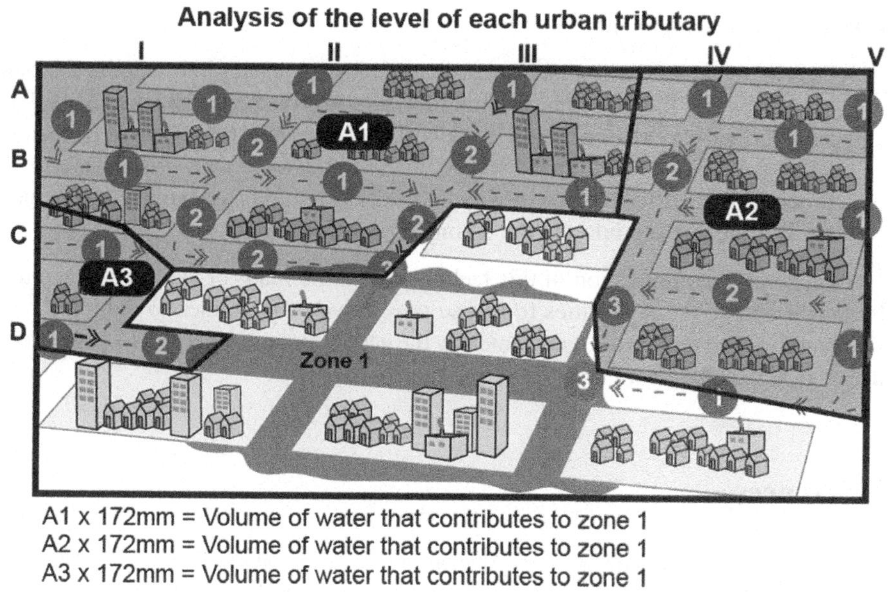

Fig. 1.13 Urban hydrographic sub-basins (Areas 1 to 3). Calculation of volume sent by each sub-basin to the flood zone (Zone 1)

1.5 Conclusion

The proposed technique may be applied in an urban area affected by a flood. It uses actual data, obtained from fieldwork involving both observations and interviews with the affected population, together with information from reports or flood warnings by the authorities. The easy-to-use technique can be used by anyone with the basic relevant knowledge, as it gives step-by-step guidance on what to do and how to do it.

Users need to refer to local urban plans, so the drawings, schemes and sketches provided are easy to understand to help to generate maps of the various levels of flood hazard in different areas. These maps crucially allow the number of lots (and even their land use) at each level of hazard level to be ascertained and, at the same time, provide estimates of the numbers of inhabitants and users of the infrastructure.

By referring to the maps or sketches generated and identifying the urban tributaries and urban sub-basins, it is possible to pinpoint those that send the greatest volume of water to the affected area, so the user has the necessary tools to:

- Develop preventive programmes in the light of knowledge of the exact number of inhabitants exposed to various levels of danger.
- Locate shelters in areas that will be unaffected by water, either surface flow or build up.
- Provide flood shelters with the materials, equipment, services and food essential to those who may be evacuated.

- Plan urban development:
  - Devise an urban development plan that avoids increasing the length of urban tributaries and therefore the volume flowing towards the flood zone, as well as not overloading existing drainage networks when connecting services from new urban areas.
  - Propose the exact site where preventive works should be constructed to reduce the volume flowing into the flood zone.

In general, the application of this technique and the maps or sketches generated by it will enable the authorities to devise preventive programmes to reduce social vulnerability and reduce the water entering the accumulation zone, and to implement urban growth strategies.

References

Alcocer-Yamanaka, V., Rodríguez, J. M., Albornoz, P. M., Maldonado, J., López, J. A., Alonso, J. A., & Quiñones, J. R. (2012). Manejo de aguas pluviales de la ciudad de Chetumal, Quintana Roo. Revista Agua en Quintana Roo, 2(4):9–14.

Bertoni, J. C. (org.) (2004). Inundaciones Urbanas en Argentina. (Ed.). GWP-SAMTAC. ISBN 987-9406-76-1. p. 270.

CENAPRED. (2006). *Basic guide for the elaboration of state and municipal Atlas of hazards and risks. Hydrometeorological phenomena*. Mexico City: Centro Nacional de Prevención de Desastres, Sistema Nacional de Protección Civil (CENAPRED).

Mobayed Khodr, N. (2009). Flood simulator in urban areas, a forecast model for warning purposes. *Science*@uaq, *2*(1), 72–90.

Mora, J., Gil, C. A., Gil, K. & López, V. (2015a). *Atlas of dangers and risks of the municipality of Tuxtla Gutiérrez, Chiapas*. SEDATU bases, internal report.

Mora, J., Gil, C. A., Gil, K. & López, V. (2015b). *Atlas of hazards and risks of the Tláhuac Delegation, Mexico City*. SEDATU bases, internal report.

Rodríguez Varela, J. M., Alcocer Yamanaka, V. H., Albornoz Góngora, P. M., Santos Téllez, R. U. & Llaguno Guilberto, O. J. (2014) *Estudio de Riesgo de inundaciones en zonas urbanas de la República Mexicana*. Jiutepec: Sectoral Fund for Water Research and Development, CONAGUA, CONACYT, IMTA.

Yamanaka, V., Rodríguez-Varela, J., Bourguett-Ortíz, V., Llaguno-Guilberto, O. (2016). Metodología para la generación de mapas de riesgo por inundación en zonas urbanas. ISSN 0187-8336. *Tecnología y Ciencias del Agua*, septiembre-octubre de 2016, 7(5):33–55.

Chapter 2
Identifying and Mapping the Risk of Rockfall and Landslide on Roads and Urban Areas

J. C. Mora Chaparro, *Mapping the Risk of Flood, Mass Movement and Local Subsidence*, SpringerBriefs in Earth Sciences, https://doi.org/10.1007/978-3-030-22472-1_2

Various terms are used to refer to slope instability in the range of disciplines involved in the study of mass movement. Ayala (2000) points out that hillside processes are referred to as gravitational processes, mass movement or, in a general sense, landslides. These terms indicate that a slope is unstable and that some form of clearance process may occur. The Spanish terminology has inherited some confusion from the English, mainly to do with the inconsistency involved in grouping several types of removal process under the umbrella term 'landslide', a term that, in both languages, means a specific, well-defined mechanism of movement; that is, a slide.

The term 'landslide' was coined by Sharpe (1938), who described it as referring to a perceptible movement downslope of a relatively dry mass of land, rock or mixture of both in a process of removal or detachment. Terzaghi (1950) explained that a removal process is the rapid displacement of a mass of rock, residual soil or hillside sediment in which the centre of gravity subsequently moves downward or outward; Varnes (1958) specified that this movement of hillside-forming material is downward, and may include natural and artificial materials. Later, Bates and Jackson (1987) defined a removal process as the transport of rocky material by the action of gravity; and more recently, in *Landslide Types and Process* (USGS 2004), explain that the term describes a wide variety of processes that result in the downward or outward movement downhill of the material that makes up a slope, including rock, soil, artificial fill or a combination of these.

2.1 Classification of Mass Movement Processes

Existing classifications of clearance processes use various parameters as their basis, for instance the type of material (rock, soil or debris), the speed (slow or fast), the water content and the break mechanism. Based on these parameters, several types may be identified.

The classification used here originated in Europe as part of a programme, 'The temporary occurrence and prognosis of landslides in the European Community', developed by the EPOCH project (1993), using work by Varnes (1958) and Hutchinson (1988).

According to this classification, the mass movement mechanisms are: landslides, rotational landslides, translational landslides and flows. These are subdivided on the basis of the material involved (rock, detritus and soil: see Table 2.1).

Table 2.1 Simplified classification of mass movement mechanism (EPOCH 1993)

Motion	Material	Speed	Effect
Creep	Soil	Slow	
Slump	Soil or debris	Slow or fast	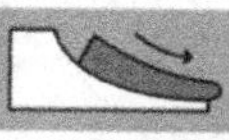
Slide	Rock or debris	Fast or slow	
Flow	Debris or mud	Fast or slow	
Avalanche	Ice and snow or debris or rock	Fast	
Fall	Rock	Fast	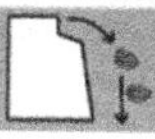

2.2 Rockfall

Falls of rock, detritus or soil can be of any size and, with an abrupt movement, may form steep gradients in both natural and artificial material. The movement is practically free-fall, rolling, bouncing or sliding downhill. A fall may be the result of erosion; detachments are generally related to structural discontinuities such as cracks, stratification planes or fracturing, which can occur on wide range of scale (Ayala 2000). The important distinction between rockfall and landslides is that in the former individual fragments move as independent rigid bodies and interact with the substrate through episodic impact (Evans and Hungr 1993).

2.3 Landslides

Slides are relatively rapid movements of a portion of a slope due to a zone of weakness that separates the most unstable material from the stable material underlying it. This surface is known as the plane of rupture, and it may be either flat or concave. Slides may be rotational or translational (USGS 2004).

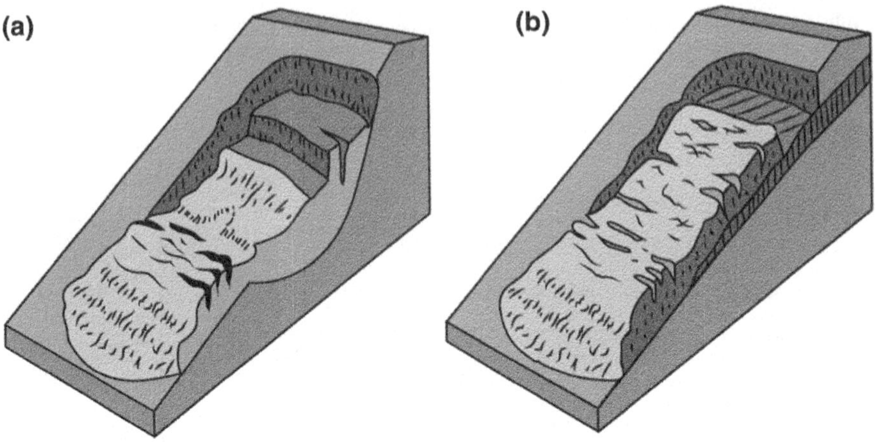

Fig. 2.1 Types of slide. **a** Rotational slide; **b** Translational slide (USGS 2004)

2.4 Rotational Slide

This type of slide has a zone of rupture that forms a concave curve, in the shape of the bowl of a spoon. The movement that occurs is rotational along the axis parallel to the surface of the ground and transverse through the slide (USGS 2004). Rotational slides tend to move at slow or moderate speeds due to self-stabilizing gravitational forces in weak rock under shear stress, which tends to fail in a ductile manner (Hungr et al. 2013). The ductile behaviour of slides is poorly understood, but is mainly a consequence of a pre-fault, in a progressive deformation that continues when the general cohesion is destroyed. The movement is active, because it has low surface tension (Hungr et al. 2013; Fig. 2.1).

2.5 Translational Slide

These landslides involve a mass of soil or rock fragments that moves outward and down a gradient as a relatively coherent mass. This motion extends along a more or less flat fault surface, with little or no rotational motion. Usually, shallow translational landslides occur in soil or are defined by weak surfaces in rock formations, such as stratification planes, zones of different alteration or meteorization of rocks in the direction of the landslide. The moving mass consists of a single unit or a few closely related units, disclosing the resistance properties of the discontinuities (Hungr et al. 2013; USGS 2004). As in rotational slides, the process leaves perceptible marks on both the original slope and the deposit site (Hutchinson 1988; Fig. 2.1).

A nomenclature to describe the parts of the land affected by a landslide has been proposed by Suárez (1998); (Fig. 2.2).

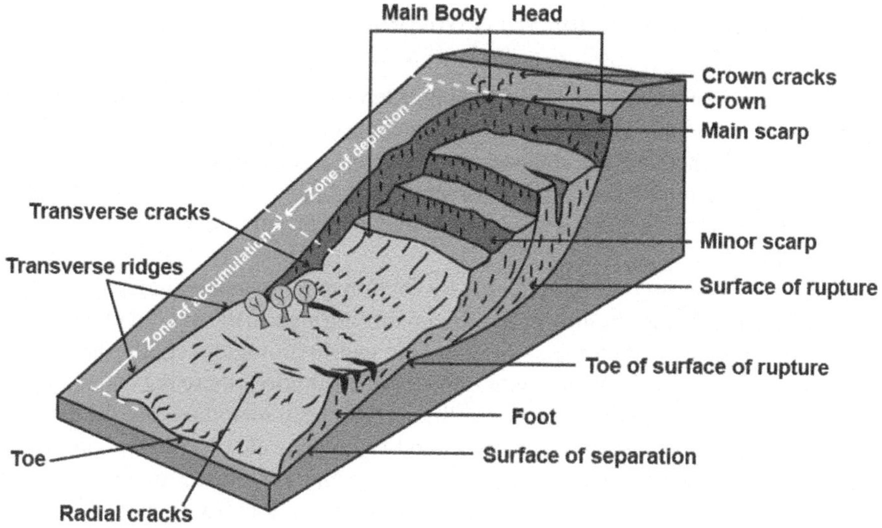

Fig. 2.2 Parts that make up a hillside affected by a mass movement (Suárez 1998)

- **Main escarpment**. A very steep surface along the periphery of the moving area, caused by the displacement of the material from the original terrain.
- **Secondary scarp**. An inclined surface produced by differential displacement within the moving mass.
- **Head**. Upper parts of the material moving along the contact between the disturbed material and the main scarp.
- **Top**. The highest point of contact between the disturbed material and the main escarpment.
- **Crown**. The material on the site that is practically unchanged, adjacent to the highest part of the main escarpment.
- **Fault surface**. Area below the movement that delimits the volume of material moved.
- **Fault surface foot**. Line of intersection between the underside of the burst surface and the original surface of the terrain.
- **Base**. The area covered by disturbed material below the foot of the fault surface.
- **Tip or nail**. The point on the base that is furthest from the top.
- **Side or flank**. The side of the movement.
- **Original surface of the land**. The surface that existed before the movement occurred.
- **Right and left**. These terms should be used looking from the crown towards the foot of the slip.

2.6 Factors Influencing Mass Movement

The occurrence of mass movement is controlled by a series of factors capable of modifying the internal and external forces acting on the ground (Terzaghi 1950). External factors increase the stresses or actions on a slope; that is, they produce a greater concentration of active motor forces. Internal factors reduce the resistance of a material; that is, they cause a decrease in the concentration of resisting forces, as shown in Fig. 2.2 (Ayala et al. 2008).

Slope instability is controlled mainly by friction, gravity and geological and geometric attributes as conditioning or passive factors intrinsic to the natural material, for instance the soil and lithological, stratigraphic and hydrological conditions, which determine the resilience and behavior of a slope. In addition, there are active factors that cause breakage once a series of conditions have been met; these are the external factors that act on soil or rocky massifs, modifying their characteristics, properties and equilibrium (Vallejo et al. 2002). If we understand how these factors affect the stability or instability of a hillside, we can devise prevention or mitigation measures (Ayala et al. 2008). In order to make susceptibility maps associated with mass movement, it is important to identify and map the factors that affect stability (Aleotti and Chowdhury 1999).

2.7 Proposed Technique

A technique is suggested for the identification and mapping of danger, vulnerability and risk of rockfall and landslide. This is a practical, simple and easy technique for gradients on roads, urban areas and riverbanks in towns and cities.

I. Identifying and mapping the hazards, vulnerability and risk of rockfall

Rockfall is a phenomenon caused primarily by gravity acting on a slope, accelerated by ground vibration, explosion, seismic activity, rain or a combination of these factors.

The technique to identify and map the risk of rockfall:

1. Identify the gradients on hillsides or in the bed and cliffs and banks alongside rivers.
2. Identify the fall hazard level; high, medium or low.
3. Map the area's various levels of danger of rockfall.
4. Record and quantify both the physical and social vulnerability to these levels of danger.
5. Assess the risk of rockfall.

2.8 Identify the Gradient and Cliff/Bank Components of a River

Rockfall refers to a block of rock changing from a state of rest to a state of vertical movement. The block can fall and move on an inclined slope, known as a bearing. These phenomena occur on gradients of over 45°, as found on roads or in urban areas and generated by their construction. Such gradients may also be found in the cliffs alongside rivers, formed by the erosive process of the water flow, and may be adjacent to urban or road development.

2.8.1 Measure the Gradient

Conduct fieldwork to identify slopes steeper than a 45° on roads or behind buildings. An approximate measure of the gradient can be achieved simply, with a protractor and ruler (Fig. 2.3).

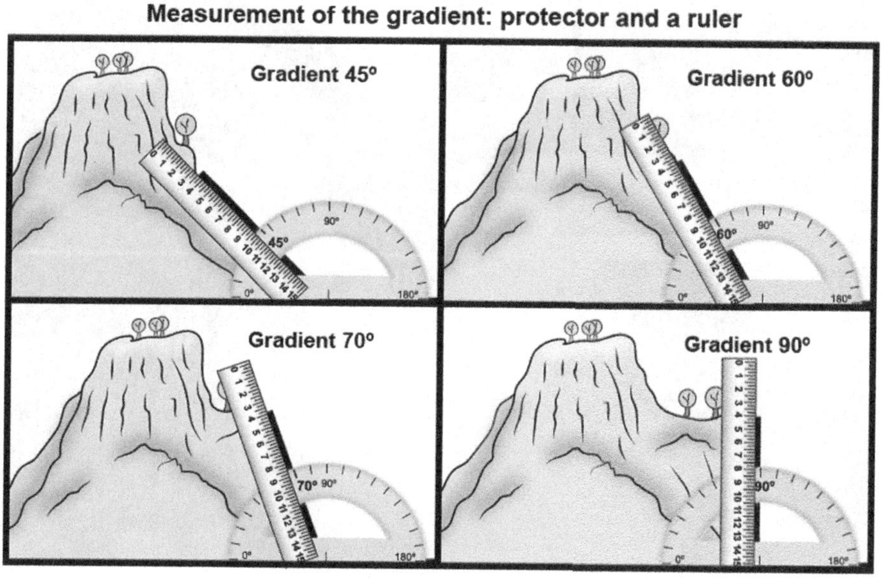

Fig. 2.3 How to measure the gradient using a protractor and ruler. The protractor should be placed along the horizontal plane of the ground. Pivot the ruler around the protractor's center-bottom mark until it lies parallel to the slope, then read the angle off the protractor

2.8.2 *Materials Making Up the Slope*

Once each gradient steeper than 45° has been located, approach as close as possible, without danger, to identify the materials that compose it: (A) just rock; (B) rock and fine material mixed; or (C) just fine material (Fig. 2.4).

(A) *Only rock. The slope may be composed of rocks of various colours and solidity. Approach it and note*:

a. The colour of the rock: light, dark or intermediate.
b. The presence of solid and hard rock that does not break under the blow of a hammer and a stable slope is observed.
c. The presence of rock that crumbles under the blow of a hammer.
d. The presence of rock with fractures. Measure the distance between one fracture and another, and record the sizes of the blocks formed by the cracks.

Fig. 2.4 Various types of rocks falling from an embankment onto road. **a** Slope with compacted rocks of dark colour, strongly fractured; fractures form blocks up to 80 cm long. **b** Road embankment with light grey compacted rock with fractures forming blocks up to 1.5 m long. **c** Slope with falls of light-coloured rocks, up to 2 m long; the rock is surrounded by fine material of the same colour. **d** Block of compacted rock of pale color, surrounded by fine material, sand and semi-compacted gravel

(B) *Rocks and fine materials mixed. Approach the slope and note*:

a. The size of the rock fragments (minimum, maximum and medium), colour, hardness if solid, or whether it crumbles easily or is fractured.
b. The material surrounding the rock. Specify the degree of compaction (compact, semi-compact or no compaction). To identify the particle size of the material, rub some between your finger and thumb and note if it is very fine (clayey), granular (sandy) or has rock fragments (gravel).

(C) *Just fine material, defined as when only one type predominates and rock pieces are rare.* Specify the degree of compaction (compact, semi-compact and/or not compacted), then take a sample in your hand to assess whether very fine material (clay), sand or gravel predominates.

2.9 Specify the Level of Danger from Falling Rocks

The rockfall hazard is assessed by direct observation of all gradients steeper than 45°.

2.9.1 High Hazard

When falling blocks may fall directly onto communication routes, infrastructure or housing (Fig. 2.5), the hazard is regarded as high.

2.9.2 Medium Hazard

When there is space between where blocks fall and the communication route, known as the gutter, buffer zone, hard shoulder, and so on, the hazard is regarded as medium. The zone should serve at least to lessen the movement of material towards the infrastructure (Fig. 2.6).

2.9.3 Low Hazard

If there is a gradient of 45° or less, there is a buffer zone and there are no blocks ready to fall although there is evidence that blocks have fallen (Fig. 2.7), it may be regarded as a low hazard.

Fig. 2.5 **a** High danger to communication route. **b** Block fall in Chiapas. **c** Diagram of falling blocks putting a house in high danger and **d** Actual rockfall near a house

Fig. 2.6 **a** The average danger of falling blocks on a communication route. **b** Falling blocks near a building under a cliff of about 6 m high, where the buffer zone is not 1:1 (6 m) but between 3 and 6 m

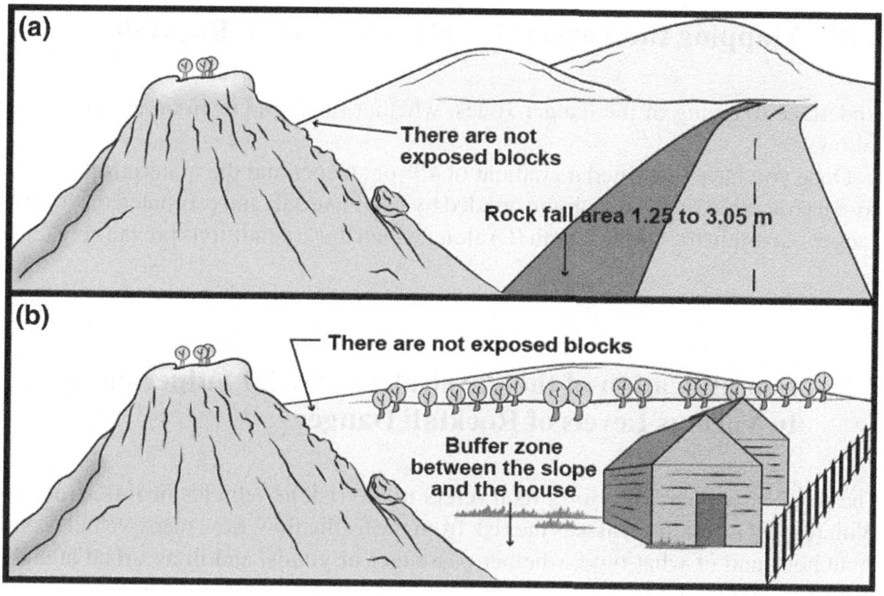

Fig. 2.7 **a** Low hazard on a road. **b** Low hazard to a building. In both cases, no blocks are about to fall, despite evidence of an earlier fall

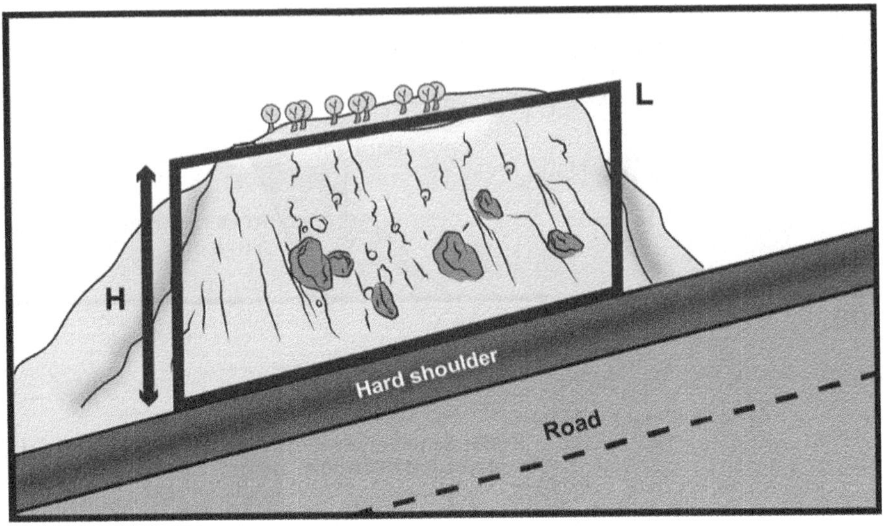

Fig. 2.8 Danger zone due to blocks falling on a road. L, length where the same conditions of slope and type of material that represent the hazard are found. H, slope height, when there are variations in height the average of the slope is obtained. Multiply H by L, A = H × L to calculate the area of the danger zone. The area may be at high, medium or low danger of rockfall. In this case, it is a medium danger area, since there is a buffer zone

2.10 Mapping the Various Levels of Danger of Rockfall

Undertake mapping of the danger zones, whether on a road or in an urban area, as follows.

Once you have identified a gradient of 45° or steeper and the materials that compose it (fractured rock, or rock surrounded by fine material) and estimated the level of danger, ascertain the whole length (L) along which these conditions prevail (Fig. 2.8).

2.11 Quantification of the Physical and Social Vulnerability to Various Levels of Rockfall Danger

The physical vulnerability of a road refers to the risk to vehicles or infrastructure. With respect to vehicles, assess the risk from the traffic flow, how many vehicles pass in an hour and of what type, whether passenger or goods, and in an urban area the structures and land use.

Assess the social vulnerability of a communication route from the number of people in each vehicle and the flows measured, and in an urban area the inhabitants and users of the different structures.

2.12 Drafting a Risk Assessment for Rockfall

Once you have identified the hazards and physical and social vulnerabilities, draw up risk assessments.

Use whatever means are available to undertake a localized assessment of each road section or urban area: if you have satnav or a map with geographical coordinates, work between specific coordinates; if you have a roadmap marked in distances, use the distance in kilometers between two towns for reference.

On each section, record the risk of falling rocks for a specific number of kilometres (km), from an average specific height (m) and across an area (m^2), involving X number of vehicles (urban transport, private, goods vehicles, etc.) and X number of people every hour. Using this information, the authorities can devise programmes of preventive works in each identified danger zone.

II. Identifying and mapping the hazards, vulnerability and risk of landslide

A slide is a mass of material that moves down a slope. When it has slipped, it is most important to traverse the entire displaced body to identify the crown, the left flank, the right flank and the base or tip, taking care. Subsequently, more detailed local identification and delimitation of the displaced corps, the active corps, the zone with open fractures and the zone with fractures should be carried out (Fig. 2.9).

The displaced body refers to the material that has changed its original position to a new one, between the crown and the base of the structure. The active body corresponds to the area of the material that has already been displaced from its original position, marked by a zone of open fractures and the base of the slide. The zone with open fractures presents a greater movement of material, developing fractures with separation (open), which indicates that it has moved. The zone with fractures is where lines may be observed in the floor yet there is no displacement. The fracture-free zone is the zone adjacent to the fractured zone.

Once you have established the various parts of a slide, identify the danger zones (Fig. 2.9).

The high danger zone corresponds to the areas of land that have already experienced displacement, and includes the zone of open fractures and the displaced corps. The medium danger zone includes that in which displacement has already occurred yet has given rise only to lines of fractures. Due to its proximity, the zone immediately adjacent to the fracture zone is regarded as a low danger zone.

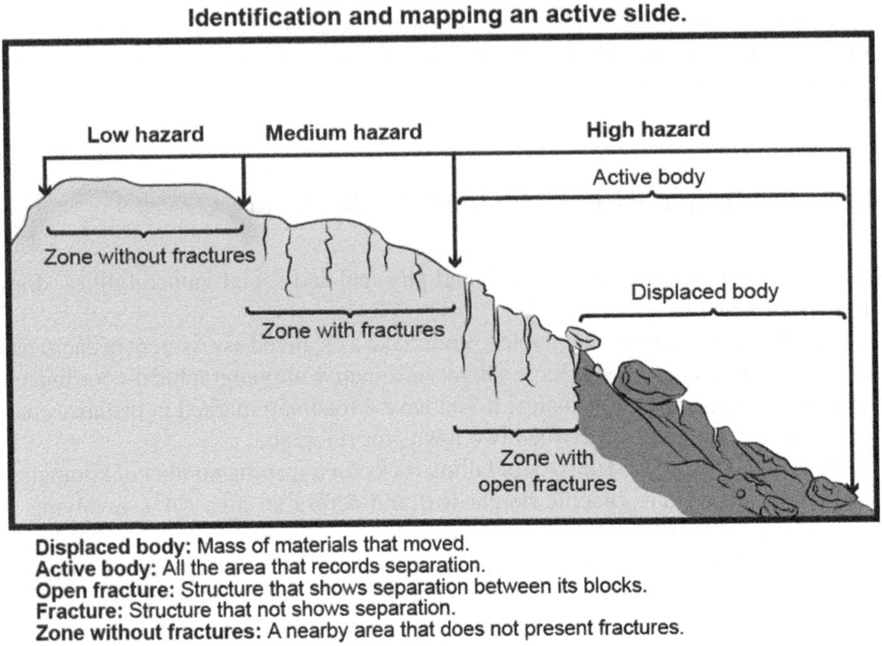

Fig. 2.9 Profile of a slide, showing the various levels of danger and the elements that comprise a sliding structure

2.13 Mapping the Danger Area in a Landslide

Undertake fieldwork to mark on a map the area of the displaced body, identifying the zones: the zone without fractures (low danger); the zone with fractures without opening (medium danger); and zone with open fractures and zone of the displaced body (high danger) (Fig. 2.10).

2.14 Physical and Social Vulnerability in a Landslide

Assess the physical vulnerability along a communication route by counting how many vehicles (vehicle flow) pass the area affected by the landslide. Assess the social vulnerability by how many people travel in each vehicle.

Assess an urban area's physical vulnerability by the risk to dwellings or common-use infrastructure in each danger zone of the landslide, and assess its social vulnerability by the risk to inhabitants or users of infrastructure.

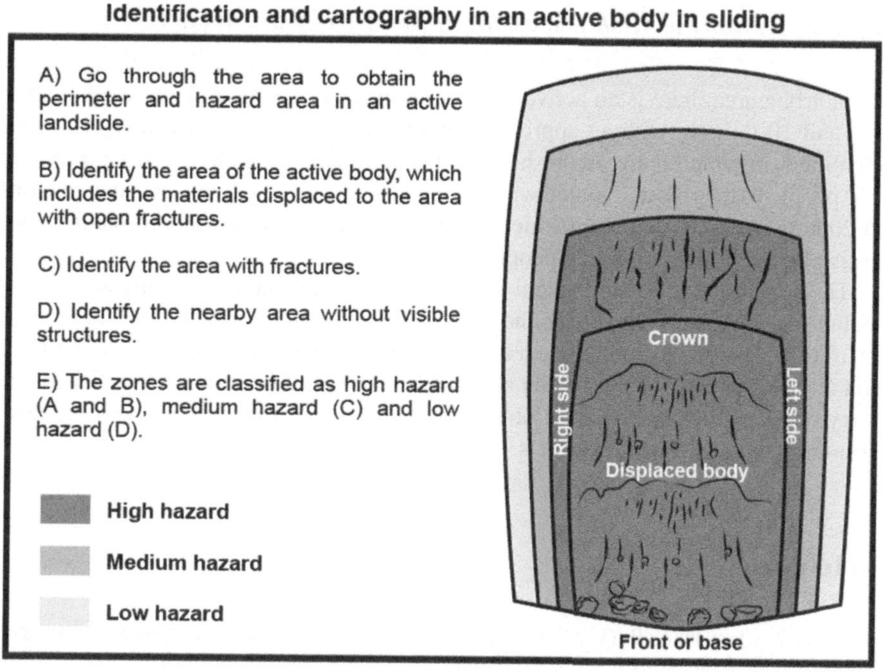

Fig. 2.10 Area of the slip structure, showing its components and hazard levels

2.15 Drafting the Slip Risk Assessment

Once you have identified the hazards and the physical and social vulnerabilities, draw up one risk assessment for each road and one for each urban area, as follows.

2.15.1 Example Risk Assessment for a Communication Route

Along the road that goes from Town A to Town B, there is a slide 350 m wide and 800 m long, on a gradient of 70° at an average height of 30 m and of an approximate area of 2800 m^2. This is bounded by an area of open fractures that has caused a displacement of the asphalt up to 5 cm, with jumps of 20 cm. The danger level is high, and 120 vehicles per hour pass, exposing an average of 360 people to risks associated with the asphalt's collapse and total displacement.

2.15.2 Example Risk Assessment for an Urban Area

In an urban area there is an active slide of 50 m wide, 300 m long, on a gradient of 60° and 10 m high, with an approximate area of 15,000 m^2. A low danger zone is identified, in which there are 10 houses, a school and a church, exposing a total of 150 people to the risk associated with the slide. There is a medium-level danger zone of 20 houses and a market, in which 200 people are at risk, and a high-danger zone of five houses inhabited by 20 people.

This technique is a fundamental tool to carry out preventive programmes, since it includes assessment of housing and infrastructure as well as the risk to inhabitants or users in a landslide zone at each level of danger. Knowing the physical conditions and all the elements of the terrain affected by the landslide, it is clearly possible to carry out more detailed studies later to put forward proposals for works to prevent or minimize the slide or its effects.

References

Aleotti, P. & Chowdhury, R. (1999). Landslide hazard assessment: Summary review and new perspectives. *Bulletin of Engineering Geology and the Environment 58*, 21. https://doi.org/10.1007/s100640050066. Berlin Heidelberg: Springer. ISSN1435-9529. Online ISSN1435-9537.

Ayala, I. (2000). *Landslides: Landslides or ground movements? Definition, classifications and terminology* (p. 9). Mexico: Investigaciones Geográficas, UNAM, Boletín del Instituto de Geografía.

Ayala, I., Salinas, C., Lopéz, M., Espinoza, J., Conde, T. &. Salvidar, S. (2008). *Inestabilidad de laderas*. Mexico: CENAPRED.

Bates, R. L., & Jackson, J. A. (1987). *Glossary of Geology*. Falls Church, VA: American Geological Institute.

EPOCH (European Community Programme). (1993). *Temporal Occurrence and Forecasting of Landslides in the European Community*. In J. C. Flageollet (Ed.), (Vol. 3), Contract no. 90 0025.

Evans, S. G., & Hungr, O. (1993). The assessment of rockfall hazards at the base of talus slopes, Canada. *Canadian Geotech Journal, 30*, 620–636.

Hungr, O., Leroueil, S., & Picarelli, L. (2013). Review, The Varnes classification of landslides types, an update. *Landslides, 11*, 67–194.

Hutchinson, J. (1988). General report morphological and geotechnical parameters of landslides in relation to geology and hydrogeology. In C. Bonnard (Ed.), *Proceedings* (Vol. 1, pp. 3–35). Rotterdam, Netherlands: A. Balkema.

Sharpe, F. (1938). *Landslides and related phenomena*. New York: Columbia University Press.

Suárez J. (1998). Deslizaminetos y estabilidad de taludes en zonas tropicales. Universidad Industrial de Santander, Bucaramanga, p. 540.

Terzaghi, K. (1950, November). Mechanisms of landslide, E.U. In *Geological Society of America, Engineering* (83–123). Geology, Berkley.

USGS. (2004). *Landslide Types and Process, E.U., The U.S. Geological Survey Landslide Program. Fact Sheet 2004–3072*. US Department of Interior Geological Survey.

Vallejo, L., Ferrer, M., Ortuño, L., & Oteo, C. (2002). *Engineering Geology*. Madrid: Pearson Education.

Varnes, D. J. (1958). Landslides types and processes. In E. B. Eckel (Ed.), *Landslides and engineering practice, Highway Res. Board Special Report* 29, 20–47.

Chapter 3
Identifying and Mapping the Risk of Local Subsidence

J. C. Mora Chaparro, *Mapping the Risk of Flood, Mass Movement and Local Subsidence*, SpringerBriefs in Earth Sciences, https://doi.org/10.1007/978-3-030-22472-1_3

3.1 Introduction

This chapter proposes a new technique to deal with subsidence in urban areas. Subsidence in several urban areas of Mexico is regarded as being closely associated with urbanization in areas located in ancient depressions filled with lake and/or fluvial-lacustrine sediment (Farina et al. 2008). At present, cities such as Querétaro, Morelia, Mexico City and suburbs, San Luis Potosí, Aguascalientes, Celaya, Guadalajara, Salamanca, Abasolo, León and Irapuato experience this phenomenon (Carreón et al. 2006).

In the case of Mexico City, subsidence in the old lake area is a major risk, manifesting itself in cracks that cause damage to municipal services and buildings. Subsidence has also had an impact on the area's seismicity (Martínez et al. 2012).

Carrillo (1948) explained the subsidence phenomenon in Mexico City and established its theoretical relationship to the consolidation of clays induced by extensive extraction of water from subsoil. Consequently, more researchers have supported his findings and contributed scientific work to providing a clear and precise solution to this problem that undoubtedly puts Mexico City at serious risk.

Other studies link the subsidence to differential sinking of land due to overexploitation of aquifers, causing surface fractures linked to geological faults (Rodríguez and Rodríguez 2006) identified by De Cserna et al. (1987). However, recent investigations by Auvinet et al. (2008) have discounted this theory, since gravimetric methods have found that they affect the basin's basement and have no superficial manifestation. The Institute of Engineering undertook a comparison of the surface fractures mapped by the UNAM geoinformatics laboratory and found that their spatial distribution bears no relation to studied seismic events. Auvinet et al. (2008) conclude that there is no relationship between 'active failures' and the generation of fractures.

Among the studies carried out in the areas affected by subsidence in Mexico City, Riquelme (1974) recorded the formation of fractures in the Tláhuac district. Santacruz (2009) analysed previous studies and concludes that the subsidence in Mixquic is associated with the lime-clay substrate found on an earlier lava flow, which could explain the vertical displacement of up to 1.66 m on Peru Street.

An unpublished report by Mora et al. (2013) as an Update to the Atlas of Risks in the Federal District found that, due to the accelerated and disorderly growth that Mexico City has been allowed, the area encompasses diverse land types, including natural zones for ponded-up water and channels, cliffs and riverbanks that are unstable to a greater extent in the rainy season. There is evidence that, in this season, these areas' streets, avenues and roads once again have ponded-up water and flowing streams and rivers.

According to the Secretariat of Civil Protection of the Federal District's (2010) stratigraphic zoning and technical regulations for the regulation of civil protection of the Federal District NTC-002-SPCDF-PV-2010, which establish technical guidelines for the approval, use, operation and operation of seismic-warning instruments in

Federal District buildings, the valley of Mexico is divided into zones: ***Lomas*** (I); ***Transición*** (II); and ***Lago*** (III), described below:

The Lomas Zone (I) is formed by rocks and generally firm soils deposited in a lacustrine environment. There can be both superficial and intercalated sandy deposits in either a loose state or cohesive yet relatively soft. It is common in this zone to find cavities in rocks, caverns and tunnels that have been excavated to exploit the sands, as well as uncontrolled landfill.

The Transition Zone (II) is constituted of volcanic materials and predominantly sandy and silty/sandy strata, interspersed with layers of lacustrine clay. The thickness of these strata is variable, between tens of centimeters and a few metres.

The Lake Zone (III) is made up of powerful, highly compressible clay deposits, with intercalations of layers of sand and layers of silt and clay. The sandy layers are generally moderately to highly compacted and of variable thickness, from centimeters to several metres. Lacustrine deposits are usually covered superficially by alluvial soils, dry material and manmade fill. The thickness of this set of sediments is 80 m in the Tarango Formation (Zeevart 1953).

Mexico's lake area was formed 500 years ago by the accumulation of sediments in what were great lakes, and consists of very soft and compressible lake deposits with a water content $n > 60\%$, with depths of up to 60 m in Texcoco and over 100 m in Tláhuac.

The technique proposed to assess the risk of local subsidence, or sinks, does not deal with the widespread subsidence caused by the regional extraction of water from the subsoil since, as its name indicates, it originates locally.

Local sinks are characterised by being preceded by structures that form little by little as a sign of a sink, collapse or vertical hollow.

The local sinks dealt with in this chapter are associated with areas of old canals, streams, rivers and drainage networks.

3.2 Proposed Technique

1. Signs of local sinks.
2. Identifying and mapping sinkhole structures related to drainage.
3. Identifying and mapping subsidence associated with local breaks in drainage pipes.
4. Identifying and mapping sinks associated with old channels or streams.

3.3 Signs of Local Sinks

Subsidence of paved streets, pavements and patios is always preceded by structures formed in accordance with the development of a vacuum structure in the subsoil. This is a hollow formed in the subsoil by the removal and thus absence of surrounding material.

When a road, pavement or playground is built, fill is first laid down and flattened, then covered by concrete or asphalt, so there is always material beneath the structure to support it. If this fill is eroded from the site, a void is formed below the matrix of asphalt or concrete that, because of its thickness and type of material, has a resistance that prevents us from seeing what is happening below. However, this resistance decreases over time, and with the passage of vehicles and people. When its resistance is reduced, small concentric or elongated cracks appear above the shaft area. These fissures follow the pattern of the internal hollow below, and its compressive strength.

If the hole below the matrix is vertical, then the fissures and fractures that form in it are concentric and semicircular or circular (Fig. 3.1). When a vertical vacuum exists or forms in the subsoil, the pressure exerted by a vehicle tyre or the weight of a person causes it to weaken and form surface fissures or concentric fractures (Fig. 3.1). These increase with further external compression and fractures and, in the centre of the structure, groundwater circulates and accumulates, further weakening the matrix and increasing the void, mainly in the centre, creating a small hole. The structure in the matrix is delimited by the vertically displaced fractures.

When the void below the matrix is inclined at an angle, the fissures or fractures in the matrix are semicircular and stepped in the direction of the angle of the void below (Fig. 3.2). As the hole increases in size in the subsoil, these structures develop and form steps descending in the direction of the slope of the hole in the matrix, through which water filters down from the surface. The structure in the matrix is delimited by fractures with a small vertical displacement and an inclined area with steps, and ends in a fracture with a larger vertical displacement (Fig. 3.2).

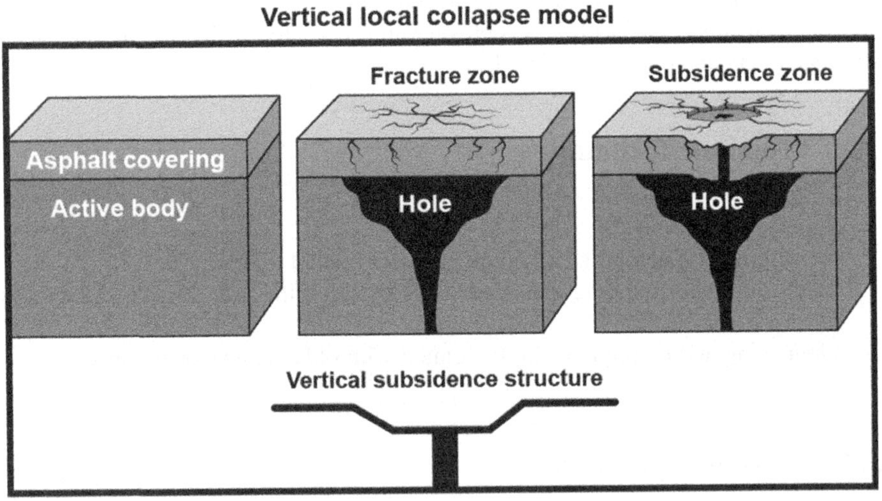

Fig. 3.1 Structures or signs associated with vertical local subsidence in asphalt or concrete

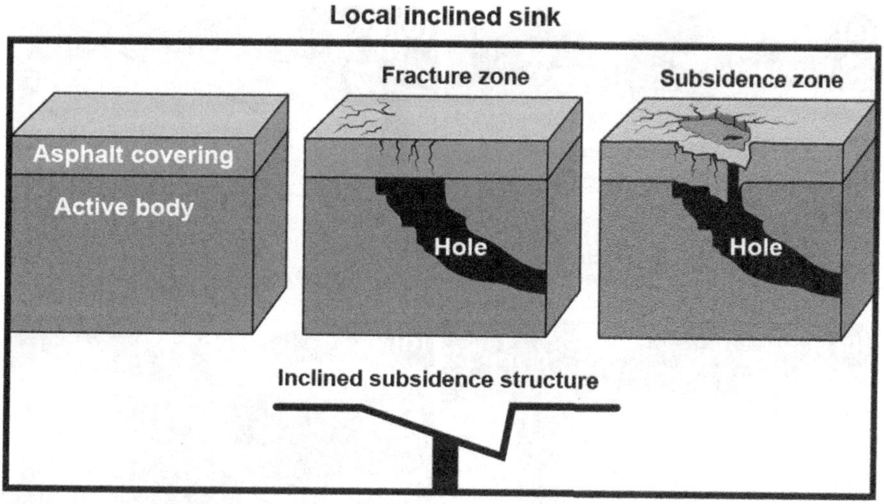

Fig. 3.2 Structures and signs associated with inclined local subsidence in asphalt or concrete

3.4 Identification and Mapping of Sinkhole Structures Related to Drainage

The structures associated with a drainage network that forms in the substrate are straight, parallel fractures and linear fractures that run down the middle of the road or over the join between the drains of an individual building and the central drainage system (Fig. 3.3).

Once the fractures have been identified, orientation data and the distance between the parallel fractures that delimit the sinking structure are recorded, as well as the height or depth of the depression in each of the fractures around the structure. Located as it is in the middle of the road, this type of structure is always related to the road drain and follows its direction (Fig. 3.3).

3.5 Identifying and Mapping Subsidence Associated with Breakage in Local Drainage Pipes

This type of subsidence is usually located in the middle of a road and is of a circular shape, with parallel, concentric fractures or small fissures (Fig. 3.4a). These structures are usually small in the asphalt layer, and larger and deeper below (Fig. 3.4b).

In this type of structure, measure the outer diameter and height (Fig. 3.4a), from the base of the inner fracture up to the asphalt (Fig. 3.4c). Since the drain's walls limit lateral transport, the structure of the subsidence is vertical and the missing material will have been eroded vertically, so it is necessary to look at the hole at depth. In

Fig. 3.3 **a** Parallel fractures in middle of road. **b** Measuring the distance between the parallel fractures that delimit the width of the structure; the drain can be seen at the centre. **c** Sinking of the asphalt layer reveals the hole beneath. **d** Recording the height and depth of each step in the hole below the asphalt

most cases, it is associated with a hole in the drain through which the material is transported by the flow down the gradient (Fig. 3.4d).

Old drain systems are subject to sinks developing under the asphalt or concrete surface; due to lack of maintenance or connection of new drains, the structure may have been weakened. When an upper wall of the pipe fractures or breaks, material is introduced from the subsoil of the surrounding matrix, forming a gap between it and the outside of the pipe. This phenomenon is not detected on the surface until cracks or fractures appear, indicating a weakening of the asphalt or concrete.

3.6 Identifying and Mapping Subsidence Associated with Old Channels of Streams

A consequence of change to land use, from agricultural and forest to urban, has long involved filling in old canals or streams with dissimilar material to create flat areas on which to construct new buildings (Fig. 3.5). The filling in of such structures has been carried out without eliminating the flow of water, so watercourses and groundwater continue to flow through the fill, causing the fine material to be eroded away. This

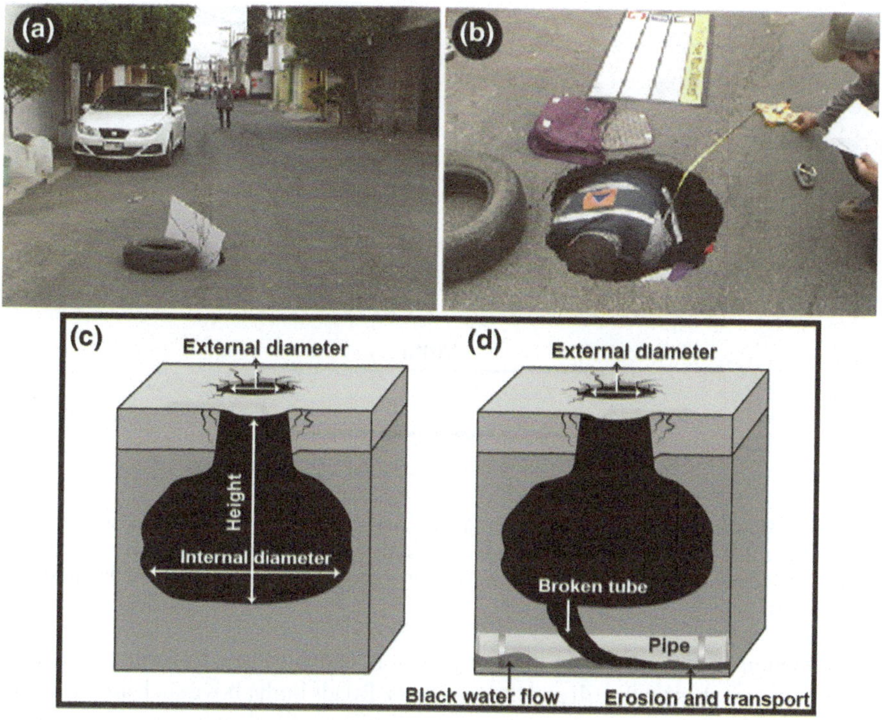

Fig. 3.4 **a** A sink in the surface of the road. **b** The great depth allows two people inside to measure the inner diameter. **c** Geometry of local subsidence. **d** How material is introduced into a hole in the wall of the pipe and water is transported in the direction of the current, which follows the gradient

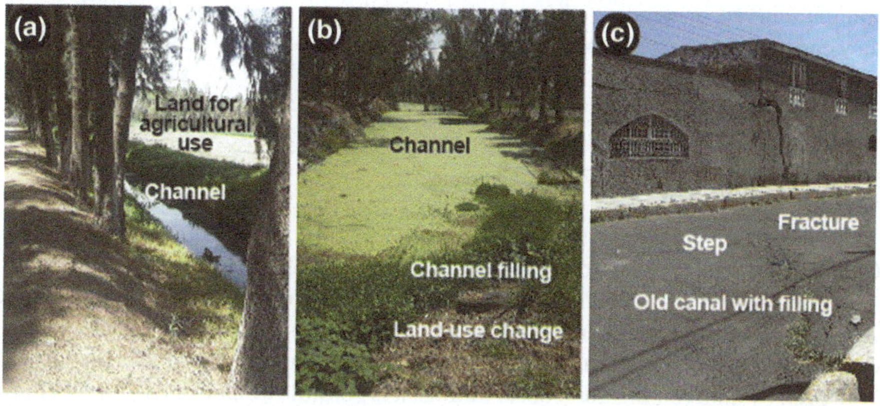

Fig. 3.5 **a** Field with channels. **b** How the canal is being filled in with various materials to become an urban area. **c** Construction on an old canal. The part over the canal is now affected by subsidence associated with erosion of material by the water still flowing in the old channel

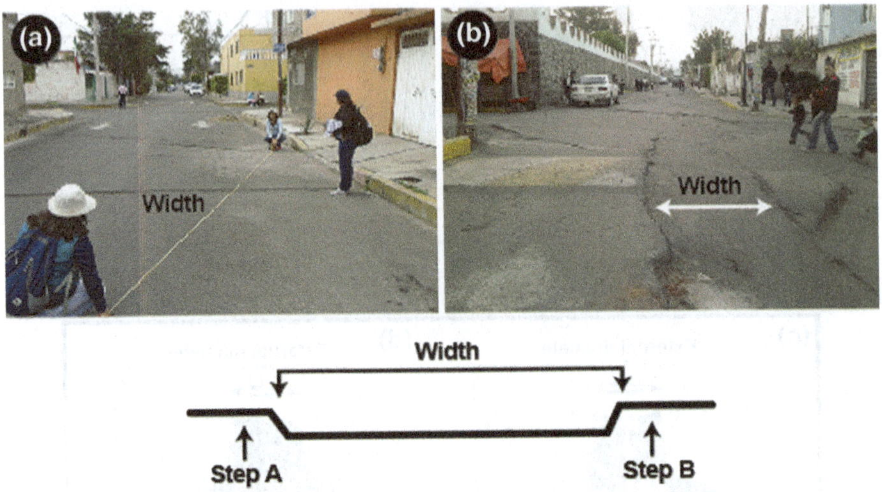

Fig. 3.6 **a** How a sink structure crosses a road. It is possible to measure the distance between the two fractures that delimit it. **b** Photograph of a sink structure first going down a road then under the wall of a cemetery. **c** Measuring the width and height of each of the steps that delimit the subsidence

phenomenon is most active in the rainy season, causing gaps to form between the material that makes up the fill and the riverbeds. Roads built above such areas begin to show linear fractures and subsidence, while dwellings or other structures experience tilting or differential movement, giving rise to fissures and fractures in walls.

The location of the structure is delimited by two parallel fractures facing each other in the asphalt or concrete matrix, showing a step downwards. These may follow the straight, curved or winding channel of the canal, river or stream. In an urban area they can lead in the direction of the road or can cross it, and thus may affect buildings to the side (Fig. 3.6).

3.7 Conclusion

Collapse associated with fractures in drainage structures and infill of old canals, streams or rivers may affect homes, buildings or roads at a local level. Such subsidence is well defined by the fractures that limit it and is thus measurable. Sometimes, it causes serious damage because it takes a long time to appear on an asphalt surface. However, by undertaking good observations, it is possible to spot signs of weakening, fractures and fissures, and ascertain whether it is an inclined or a vertical collapse.

One of the first questions to ask to identify these sinks is where the material went that had been there. The difference between deformation of the matrix and signs of subsidence is that, in the case of deformation, the material from the subsoil will have accumulated in front of or next to the structure, whereas in the case of subsidence due to a hole in the subsoil, it will not; the subsidence will not have deformed the ground, just caused a variation in height. The most important exercise in identifying the cause of a sinkhole is to survey local inhabitants to establish the land use prior to it becoming a road or a buildings being constructed. In most cases, their answer will be more accurate than any study.

References

Auvinet, E., Reverest, L., St-Arnaud, A., Rosseau, J., Meunier, J. (2008). Fall detection using multiple cameras. In EMBS 2008, 30th Annual International Conference of the IEEE.

Carreón, D., Hidalgo, C., & Hernández, M. (2006). Fracking mechanisms of clay deposits in urban areas. Differential deformation case in Chalco, State of Mexico. Bulletin of the Mexican Geological Society. *Special Issue Urban Geology, LVII*(2), 237–250.

Carrillo, N. (1948). Influence of artesian wells in the sinking of Mexico City. In *Proceedings 2nd international conference on soils mechanics* (p. 7). Rotterdam, Netherlands: International Society for Soil Mechanics and Geotechnical Engineering.

De Cserna, S., De la Fuente, M., Palacios, M., Triay, L., Mitre-Salazar, L., & Mota, R. (1987). Geological structure, gravimetry, seismicity and regional neotectonic relations of the Mexican basin. *Universidad Nacional Autónoma de México, Instituto de Geología, Boletín, 104,* 71.

Farina, P., Avila-Olivera, J.A., Garduño-Monroy, V.H. and Catani, F. (2008). SinSAR analysis of differential ground subsidence affecting urban áreas along the Mexican Volcanic Belt (MVB). *Rivista Italiana di Telerilevamento, 40*(2), 103–113.

Martínez, J., Lermo, J., Sanchez, F., Angulo, J., Valle, R, Ordoñez, J. & Perez, L. (2012). Effects of the subsidence on the changes of dominant periods of soils within Mexico City valley. Abstract presented in *15th World conference on earthquake engineering,* 22–28 September, Lisbon.

Mora, J., Gil, A., López, V. & Ramírez, O. (2013). *Atlas de Riesgos de la Delegación Tláhuac* (unpublished).

Riquelme, D. (1974). Fracturas del fraccionamiento Los Olivos en la región cercana al volcán San Lorenzo o San Nicolás, Tláhuac. *Yearbook of Geography, 14,* 67–72.

Rodríguez, R., & Rodríguez, I. (2006). Consecuencias sociales de un desastre inducido, subsidencia. *Boletín de la Sociedad Geológica Mexicana, 2,* 265–269.

Santacruz, B. R. (2009). Causas y consecuencias de la generación de hundimientos en San Andrés Mixquic, D.F. Tesis Ingeniería, *Universidad Nacional Autónoma de México*, p. 120.

Zeevaert, L. (1953). Outline of the stratigraphical and mechanical characteristics of the unconsolidated sedimentary deposits in the basin of the Valley of Mexico. In *Proceedings of the 4th congress international association quarterly result (INQUA)* (pp.1–14).

Conclusion

Collapse associated with fractures in drainage structures and infill of old canals, streams or rivers may affect homes, buildings or roads at a local level. Such subsidence is well defined by the fractures that limit it and is thus measurable. Sometimes, it causes serious damage because it takes a long time to appear on an asphalt surface. However, by undertaking good observations, it is possible to spot signs of weakening, fractures and fissures, and ascertain whether it is an inclined or a vertical collapse.

One of the first questions to ask to identify these sinks is where the material went that had been there. The difference between deformation of the matrix and signs of subsidence is that, in the case of deformation, the material from the subsoil will have accumulated in front of or next to the structure, whereas in the case of subsidence due to a hole in the subsoil, it will not; the subsidence will not have deformed the ground, just caused a variation in height. The most important exercise in identifying the cause of a sinkhole is to survey local inhabitants to establish the land use prior to it becoming a road or a buildings being constructed. In most cases, their answer will be more accurate than any study.

J. C. Mora Chaparro, *Mapping the Risk of Flood, Mass Movement and Local Subsidence*, SpringerBriefs in Earth Sciences, https://doi.org/10.1007/978-3-030-22472-1

Glossary of Terms

Altitude The height of a geographical point, measured from mean sea level [4].

Artesian The level of water accumulated in the subsoil on an impermeable layer of the ground; it can be used by means of wells. This name is also given to the layer of the subsoil that contains and stores them [4].

Asphalt It is a solid and compact mixture of hydrocarbons and minerals that is used mostly to build the pavement of roads [10].

Basin The totality of the area drained by a river or its tributary, such that all the natural run-off originating in the area is discharged through a single outlet [2].

Channel The bed of a river or stream along which water flows [3].

Collapse A geological phenomenon that consists of the abruptly free-falling and rolling of material, from vertical or almost vertical clefts in uneven terrain. It differs from slides, as free fall is the main form of movement and there is no well-defined sliding surface. Landslides can involve both rock and soil. Soil collapses are generally not of great magnitude, as the lack of consolidation prevents the formation of deep clefts; on the other hand; rock collapses can occur in cliffs and uneven terrain [4].

Collapse or settlement It is the settlement under the material that covers a surface, this is carried out by a slow movement of horizontal form under that cover, not in a free surface of a rigid cover (Thornbury 1966). The most common cause is the slow removal of material below the deck that will sink [2].

Compaction The decrease in the thickness of a stratigraphic sequence due to the weight and pressure of overlying rocks (lithostatic pressure) [6].

Contour A continuous line describing the relief of the surface. All points on a contour are at the same elevation [3].

Cracks Shallow fracture of a rock, soil or mineral [6]. Synonym for joint, diaclase or fracture [6].

Damage The loss or impairment suffered by people or infrastructure, as a consequence of environmental change, understood as damage to health, death or any other negative effect caused directly or indirectly by exposure to materials or waste, or damage to the environment [2].

J. C. Mora Chaparro, *Mapping the Risk of Flood, Mass Movement and Local Subsidence*, SpringerBriefs in Earth Sciences, https://doi.org/10.1007/978-3-030-22472-1

Danger The probability of potentially harmful natural or technological phenomenon, for a specific period and in a specific locality. It is identified, in most cases, with the support of science and technology [2].

Danger Assessment of the maximum expected impact of a destructive event in a given area and over a given period, based on probability analysis [2].

Deforestation Loss of natural vegetation in a geographical region as a result of human activity [1].

Desiccation or contraction crack Crack formed in fine material, sludge, mud and clay that, after being saturated with water and dried by an increase in temperature. It has a characteristic hexagonal shape [6].

Drainage network The drainage network of a basin comprises a main channel and a series of tributaries from the higher parts [3].

Erosion The removal of soil and rock particles by wind, water and ice [1].

Escarpment Abrupt slope, of variable height, which can be formed by tectonic movement and exogenous processes, but mainly gravitational processes [7].

Fall Fragments of soil or rock that fall off and are deposited on the lower part of a slope [1].

Fluvial erosion Destruction of rocks by fluvial processes which, together with gravitational movements, leads to the formation of valleys or lowering of the surface. The process also includes the mechanical destruction of the rocks, the washing and lamination of river valleys and the chemical alteration of rocks [2].

Fluvial process Processes that modify the earth's surface by means of line temporary and permanent and surface run-off. They cause forms of relief through erosion and ponding [7].

Fracture Rupture of rock without displacement of the separated blocks. It is also a synonym for 'fault' [2].

Fragment A piece of rock, classified according to size into: blocks, cobbles, sands, silts and clays [6.

Groundwater Water below the ground's surface [1].

Heave Slow movement, sometimes practically imperceptible, depending on the slope of a part of the natural slope, involving a mass of soil or detritic material (rocks formed by fragments or detritus coming from the erosion of pre-existing rocks). The movement is not homogeneous and within the mass several partial movements can be distinguished. The vertical displacement is a few centimeters and the horizontal is almost null, being this the characteristic that differentiates it from a slide.

Hydrological cycle The different phases or stages through which water passes as it moves across the earth's surface:

(a) Formation of water masses (oceans, seas, lakes, lagoons, etc.)
(b) Evaporation and transpiration
(c) Formation condensation and saturation of clouds
(d) Rain
(e) Infiltration and formation of groundwater, where the temperature is low and precipitation is in the form of snow or hail, forming ice deposits or glaciers. The process is repeated indefinitely, giving rise to hydrological processes [6].

Hydrography The study of the physical, chemical, biological, static and dynamic aspects of the Earth's water [4].

Hydrology The study of the presence and movement of water, both underground and surface [1].

Hydrological regionalization Procedures that estimate a hydrological variable (usually the flow) of a site where there is little or no data from other sites [2].

Hillside Side of hill or a mountain [1].

Instability of natural slope Also known as a landslide, it involves movement of rocks and soil by the action of gravity. Past landslides have been responsible for the topographical features of today's natural landscape [1].

Isoyets Lines joining points of equal precipitation [1].

Lithology The stratigraphic characteristics of a geological formation or an area of land, i.e. the types of rock, grain size, colour and mineral constituents [1].

Mitigation (reduction) Planning and execution of intervention measures aimed at reducing or diminishing risk. Mitigation is the result of accepting that it is not possible to control risk completely; that is, in many cases it is not possible to prevent or avoid damage and its consequences completely, only to lessen them [3].

Natural slope Mountain or hillside [1].

Prevention Measures and actions taken in advance in order to prevent the occurrence of a dangerous phenomenon or to reduce its impact on the population, goods and services and the environment. One of the stages is the disaster cycle, which consists of preventing the event from occurring, recognizing that it is sometimes impossible to avoid such an occurrence [3].

Regional sinking and cracking Phenomena of a geological nature whose presence is due to soft soils, in which volume losses occur as a result of the extraction of water from the subsoil [4].

Recurrence interval The average time, expressed in years, that must elapse for an event to occur in which a threshold is exceeded [2].

Risk Probability of damage caused by a disturbing phenomenon (General Law on Civil Protection). UNESCO defines risk as the possibility of loss of life, property or production capacity. This definition involves three aspects related by the following formula: risk = vulnerability × value × hazard. In this relationship, value refers to the number of human lives threatened or, in general, to any of the economic elements (capital, investment, productive capacity, etc.) exposed to a destructive event. Vulnerability is a measure of the percentage of the value that can be lost in the event of a given destructive event. The last aspect is the probability that a particular area will be affected by some of the destructive manifestations of calamity [2].

Sedimentary deposit Material accumulated in depressions in the earth's crust, land or water, carried by agents such as: rivers, wind, seas, lakes, gravity and chemical action [6].

Seismic regionalization Terrestrial zoning of a determined region and characterized by the occurrence of earthquakes, differentiating one zone from another by its greater or lesser intensity [4].

Sinking Dislocation of the earth's crust that results in the vertical removal of fragments of the crust [4].

Slip Applied to soil and surface material, refers to the slow downward plastic movement on a sliding plane [1].

Slope Slope formed by the accumulation of rock fragments at the foot of cliffs or mountains. The rock fragments that form the slope can be debris, sliding material or broken pieces detached by the action of frost. However, the term slope is actually very widely used to refer to the rock rubble itself.
The generic name of slope refers to anybody of earth and/or rocks, which are delimited by an inclined surface and form a determined angle with respect to the horizontal. Slopes are classified as natural and artificial.
When the slope occurs spontaneously, according to the laws of nature (without human intervention), it is called natural slope or simply slope.
When man does it, it is called an artificial slope, which can be a cut or an embankment, or simply a slope. In order to make a cut, the excavation is carried out in one or more geological formations, while the artificial slopes are the inclined sides of the embankments built with selected materials and mechanically compacted.

Soil A solid porous structure of heterogeneous composition that occupies the most superficial part of the lithosphere. The soil not only supports organisms, but also contains water and nutrients.

Surface run-off Part of the precipitation flowing through the surface of the soil [2].

Underground water Water beneath the surface of the ground; also referred to as groundwater [1].

Vulnerability It is defined as the susceptibility or propensity of systems exposed to being affected or damaged by the effect of a disturbing system; .e. the degree of expected losses; ease with which a system can change its normal state to one of disaster, by the impacts of a calamity [2].

Unstable floor A geological phenomenon also known as earth movement, which consists of the downward movement of soil and rock on land with slopes or unevenness, caused by the gravitational thrust of its own weight, when it overcomes the opposing forces of friction, material cohesion, or vertical or lateral containment [4].

Water table Highest surface area of the groundwater saturation zone. It is irregular, and its gradient and shape are determined by the amount of groundwater and by the permeability of the rock [1].

Watershed An imaginary line formed by points at the highest topographic level, delimiting a basin and separating it from neighbouring basins [3].

Bibliography

CENAPRED Glossary. (2010). http://www.cenapred.unam.mx/es/Glosario/. Online consultation. Accessed January10, 2017.

Bases for Standardization in the Elaboration of Risk Atlas and Catalog of Geographic Data to Represent Risk. (2014). Secretariat of Territorial and Urban Agrarian Development Undersecretariat of Territorial Ordinance. General Directorate of Territorial Planning and Risk Zone Attention (Glossary).

Guía técnica para la incorporación del Análisis de Riesgo en los Ordenamientos Ecológicos Municipales y Regionales, (2009) Final Report, Cuauhtémoc León.

Civil Protection, Secretaría de Gobernación (online consultation). http://www.proteccioncivil.gob.mx. Accessed January 10, 2015.

INEGI. Guía para la Interpretación de Cartografía Edafología, http://www.inegi.org.mx. (PDF online). Accessed January 10, 2015.

edu academy. *Geological Dictionary* (online consultation). http://www.academia.edu/4059077/DICCIONARIO_GEOLOGICO. Accessed January 10, 2015.

Lugo H, J. I., (2008). *Geomorphological Dictionary.* Institute of Geography. Mexico: UNAM.

Strahler, A., & Strahler, A. (1989). *Physical Geography.* Barcelona: Ediciones Omega. S.A.

Tarbuck, & Lutgens. (2005). *Earth Sciences. Introduction to Physical Geology.* Madrid: Pearson-Prentice Hall.

Wordreference. (n.d.). Retrieved 22 January 2015 from http://www.wordreference.com/definicion/asphalt.

Definition ABC. (n.d.). Retrieved 22 January 2015 from http://www.definicionabc.com/geografia/geofisica.php.

Project builder. (n.d.). Retrieved 22 January 2015 from http://constructoracmproyectos.com/que-es-mamposteria-en-construccion/.

Definition de punctual. (n.d.). Retrieved 22 January 2015 from http://definicion.de/puntal.

Agiusa. (n.d.). Retrieved January 22, 2015 from https://www.agiusa.com/ministing.sp.shtml.

Lakubovskii, I., & Liakhov, L. (1980). *Electric Exploration.* Moscow: Reverte.

Davila, B. (2011). *Geological Dictionary.* Peru: INGEMMET.

Norma técnica complementaria al reglamento de la Ley de Protección Civil del Distrito Federal. http://cgservicios.df.gob.mx/prontuario/vigente/3425.htm. Accessed November 23, 2018.

Dictionary of de concepts about natural sciences. https://deconceptos.com/ciencias-naturales/corrosion#ixzz3RvyoLBvD. Accessed September 12, 2018.

J. C. Mora Chaparro, *Mapping the Risk of Flood, Mass Movement and Local Subsidence*, SpringerBriefs in Earth Sciences, https://doi.org/10.1007/978-3-030-22472-1

Uncited References

Alcocer Yamanaka, V. H., Rodríguez Varela, J. M., Albornoz Góngora, P. M., Maldonado, S. J., López Orozco, J. A., Alonso Ovando, J. A., et al. (2012). Integral stor mwater management. *Agua QRoo Magazine*, 8.

Alcocer Yamanaka, V. H. et al. (2016). Methodology for the generation of flood risk maps in urban areas. *Technol. Cienc. Water* [online, 7, 5 33–55 [cited 2017-12-30], Available: http://www.scielo.org.mx/scielo.php?script=sci_arttext&pid=S2007-24222016000500033&lng=en&nrm=iso. ISSN 2007-2422.

Bertoni, J.C., Ambrosio, S., Barbeiro, O., Daniels, A., Maza, J. A., Paoli, C. U. & Serra, J. J. (2004). *Inundaciones urbanas en Argentina*. Universitas Córdoba: Editorial Científica Universitaria, ISBN: 987-9406-82–6.

Domínguez, M. R. (2017). Las Inundaciones en la Ciudad de México. Problematic and Alternative Solutions. *Revista Digital Universitaria* 1 (2). http://www.revista.unam.mx/vol.1/num2/proyec1/. Accessed 02/01/2018.

Juan Carlos Mora Chaparro (2011). Atlas de Riesgo del Municipio de Suchiate. SEDATU, p. 130. http://rmgir.proyectomesoamerica.org/PDFMunicipales/2011/vr_07087_AR_SUCHIATE.pdf.

Juan Carlos Mora Chaparro et al., (2011). Atlas de Riesgo del Municipio dc Tapachula. SEDATU, p. 143. https://www.academia.edu/36078309/ESTE_Atlas_de_Riesgos_Naturales_del_Municipio_de_Tapachula_2011.

Juan Carlos Mora Chaparro (2011). Atlas de Riesgo Delegación Milpa Alta, DF. SEDATU, p. 102. http://www.normateca.sedesol.gob.mx/work/models/SEDESOL/Resource/2612/Atlas_Estados/09009_MILPA_ALTA/ATLAS_RIESGOS_MILPA_ALTA.pdf.

Juan Carlos Mora Chaparro (2013). Sistema Integral de protección Civil Estado de Chiapas. Fase 1 y 2. Duración 2011 hasta 2013. Fecha de inicio: 2011, Fecha de conclusión: 2013. Atlas de riesgos estatal digital, reporte interno. http://www.sistemaintegraldeproteccioncivil.mx/.

Juan Carlos Mora Chaparro (2014). Atlas de Riesgos de la Delegación Xochimilco. SEDATU. http://www.atlas.cdmx.gob.mx/mapas/MR_Xochimilco.pdf. Juan Carlos Mora Chaparro Actualización de Atlas de Riesgos de Tuxtla Gutiérrez, Chiapas. Municipio de Tuxtla Gutiérrez, Chiapas. 324p. https://tuxtla.gob.mx/Atlas-de-riesgos-actualizacion-2015.

Juan Carlos Mora Chaparro (2014). Atlas de Riesgos de Tlahuac. Distrito Federal, p. 154. http://www.atlas.cdmx.gob.mx/mapas/MR_Tlahuac.pdf.

Marsal, R.J., M. Mazarí, (1959) The subsoil of Mexico City. Abstract presented at the *First Panamerican Conference on Soil Mechanics and Foundation Engineering*, Mexico City, 7–12 September.

Rodríguez Varela, J. M., Alcocer Yamanaka, V. H., Albornoz Góngora, P. M., Llaguno Guilberto, O. J., Maldonado, S. J. (2014) *Problemática de inundaciones en zonas urbanas y propuestas de solución desde un enfoque matemático*.https://www.imta.gob.mx/problematica-de-inundaciones-en-zonas-urbanas-y-propuestas-de-solucion-desde-un-enfoque-matematico. Accessed online on December 30, 2017.

Secretariat of Civil Protection of the Federal District. (2010). Stratigraphic zoning and technical regulation complementary to the regulation of the law of civil protection of the Federal District NTC-002-SPCDF-PV-2010 that establishes the technical guidelines for the approval, use, operation and operation of seismic warning instruments in buildings of the Federal District. Secretariat of Civil Protection of the Federal District (Internal document, unpublished).

Tláhuac Urban and Territorial Development Program (PDDU). (2008). *Tláhuac Urban and Territorial Development Program*. Legislative Assembly of the Federal District, IV Legislature.

The manufacturer's authorised representative in the EU is Springer Nature Customer Service Centre GmbH, Europaplatz 3, 69115 Heidelberg, Germany. If you have any concerns regarding our products, please contact ProductSafety@springernature.com

Printed and bound by CPI Group (UK) Ltd, Croydon, CR0 4YY
05/10/2026
02228152-0003